Multidimensional Analysis and Discrete Models

Aleksei A. Dezin
Steklov Mathematical Institute
Russian Academy of Sciences
Moscow

Translated from the Russian by Irene Aleksanova

CRC Press
Boca Raton New York London Tokyo

Library of Congress Cataloging-in-Publication Data

Dezin, A. A. (Alekseĭ Alekseevich)
[Mnogomernyĭ analiz i diskretnye modeli. English]
Multidimensional analysis and discrete models / A.A. Dezin ;
translated from the Russian by Irene Aleksanova.
p. cm.
Includes bibliographical references and index.
ISBN 0-8493-9425-2 (alk. paper)
1. Multivariate analysis. 2. Mathematical models. I. Title.
QA278.D4413 1995
530.1'5—dc20

95-16807
CIP

Direct all inquiries to CRC Press, Inc., 2000 Corporate Blvd., N.W., Boca Raton, Florida 33431.

International Standard Book Number 0-8493-9425-2
Library of Congress Card Number 95-16807
Printed in the United States of America 1 2 3 4 5 6 7 8 9 0
Printed on acid-free paper

Preface

During the last decades, much research effort was devoted to the study of topological, geometrical, and algebraic structures underlying the basic ideas of classical analysis. The results obtained made it possible, in a number of cases, to construct certain intrinsically defined discrete models (finite or "countable") for the problems of analysis and mathematical physics. It is to be understood, however, that when speaking about these models we do not mean just approximation (in one way or another) of the given continual object. What we do mean is the construction of discrete counterparts of these objects.

It should also be mentioned that, recently, similar ideas have started to penetrate computational practice. Treatment of discrete models is of some methodological interest as well since it allows us to acquire a better understanding of the nature of this or that relation or statement.

Formally, the main subject matter of the book is the description of some regular methods of constructing intrinsically defined discrete models for special classes of continual objects.

In fact, I had a more extensive task in mind. I wanted (1) to demonstrate (and do it at as "low" a level as possible) the interaction of ideas and methods of such parts of mathematics as, for example, classical and functional analysis, Riemannian geometry, and algebraic topology; and (2) to demonstrate this interaction not when studying such a special and difficult problem as the index of the general elliptical operator on a smooth manifold, but when investigating the connection between the Laplace operator, the multidimensional analogs of Cauchy–Riemann equations (or operations of vector analysis), and the corresponding difference equations.

The book is addressed, first of all, to those who have chosen mathematical physics as their main field of interest. It is assumed that the potential readers would like to acquire a more intimate knowledge of the nature of some foreign objects they have to deal with in the course of their research, without resorting to the study of fundamental works devoted to Riemannian geometry, geometrical theory of integration, algebraic topology and the like, which they have no wish, time, or energy to pursue systematically. It is also my belief that, in a number of cases, the study of special finite models can be very useful and can simplify the acquaintance with some related divisions of mathematics.

This approach has considerably influenced the treatment of the material in the book. The reader is supposed to be more or less familiar with partial differential equations, to have some notion of differential geometry, and to have never been concerned with problems or definitions of homology theory.

I have few words about the arrangement of the material. The introductory chapter should give a fairly clear idea of what is meant by a finite model and Chapter 1 of how the multidimensional analysis should be understood. In Chapter 2, the reader will see (perhaps not without a certain surprise or even disappointment) that analysis on the Riemannian manifold is treated as the theory of special class of the first-order partial differential equations closely related to the operators of classical vector analysis, the Laplace operator, and wave operator. Such concepts as curvature, connectedness, G-structures, and the like are not mentioned at all.

Chapters 3–5 are devoted to models proper. Chapter 3 considers the objects of classical mathematical physics, Chapter 4 deals with quantum mechanics and the field theory, and Chapter 5 considers some general aspects of the theory of discrete equations which do not claim to be connected to physics in any way.

For a sufficiently advanced reader, all chapters are practically self-contained (modulo necessary notations and definitions).

A sufficiently complete formal outline of the subject matter is given in the table of contents. More detailed and less formal information can be found in the introductions and preliminary notes to each chapter and section.

Contents

To the Reader

The book is divided into chapters, chapters into sections, and sections into subsections. The numbering of formulas, theorems, and statements is its own within every section. When a relation or theorem is cited in the framework of a given section, only its number is indicated, and when a relation or theorem from some other section is cited, that section or subsection is indicated in addition to the number. When necessary, a chapter is also indicated.

A number in brackets means a reference to the corresponding number in the list of references. The reference does not mean that the indicated book or article is the only source (or the initial source) of the revelant information.

The use of Halmos' sign ■ meaning the end of a proof (maybe only outlined) or emphasizing the absence of a proof is not completely formalized. In certain cases it is omitted.

As a rule, definitions are not separated as a special paragraph but are usually entered into the main text. Concepts being defined are given in italics. Elements of the text which are specially emphasized are given in bold face.

Chapter 0

One-Dimensional Models

0. Introductory Remarks

As was noted in the Preface, this chapter must give an idea of the most important features of models considered in the main part of the book. Models on the real line correspond to those considered in Ch. 3 and models on a circle, to those from Ch. 4.

At the same time, one more objective is pursued. It is known that when analogs are studied and used (in this case the analogs between objects of continual and discrete nature), we proceed from a thorough analysis of axiomatic chains (chains of definitions) that describe the objects being compared. Definitions occupy a considerable place in this exposition. A direct description and sufficiently detailed consideration of discrete structures that simulate (from the chosen viewpoint) the analysis on the real line and on a circle must convince us of the necessity to study attentively the formal definitions (and the nonformal additional remarks) given in Ch. 1.

It stands to reason that in this "direct description" we inevitably find a number of standard concepts which we again discuss in Ch. 1. At the same time, the relationship between the nonstandard objects and the more sophisticated formalism is later indicated as far as practical.

It should also be pointed out that the arguments presented in 1.3, which refer to the elementary second-order equation, may seem to be too complicated, and duplicate the scheme for studying the Poisson equation in Sec. 3, Ch. 3.

1. Models on the Real Line

1.1. Combinatorial real line. We introduce a certain object which is a combinatorial model of an oriented real line, for which purpose we divide by points the oriented real line into an infinite system of intervals each of which is finite. We consider the partitioning to be regular, i.e., the points that define it cannot merge. The absence of any scale, i.e., the arbitrariness of the lengths of the intervals, must be emphasized. Having chosen the base point x_0, we denote the next point on the right by x_1, the next point on the left by x_{-1}, and so on, numbering all the points by a system of integer-valued indices $k = 0, \pm 1, \pm 2, \ldots$. It is convenient to introduce shift operators τ, σ,

$$\tau k = k + 1, \qquad \sigma k = k - 1,$$

in the set of indices. We denote the open interval $(x_k, x_{\tau k})$ by e_k. We shall regard the sets $\{x_k\}$, $\{e_k\}$ as sets of base elements of the real linear spaces $\mathfrak{C}^0$, $\mathfrak{C}^1$, i.e., consider the linear combinations

$$a = \sum a^k x_k, \qquad b = \sum b^k e_k, \tag{1}$$

which are customarily called chains (of dimensions 0 and 1, respectively), to have sense. When writing a sum of form (1), we suppose, unless otherwise specified, that only a finite number of terms are nonzero. We denote by $\mathfrak{C} = \mathfrak{C}^0 \oplus \mathfrak{C}^1$ the direct sum of the introduced spaces. We define the boundary operator, namely, the mapping ∂, in $\mathfrak{C}$,

$$\partial e_k = x_{\tau k} - x_k, \qquad \partial x_k = 0, \tag{2}$$

which associates the oriented interval with its boundary and annuls the points. The definition is linearly extended to chains (1). We call the object $\mathfrak{C}$ that we have introduced a *combinatorial real line.*

REMARK. If we consider the definitions given in Sec. 2, Ch. 1, we see that we have defined $\mathfrak{C}$ as an elementary infinite one-dimensional complex. However, for the time being, we do not use this term.

Let us now consider the concept of functions over $\mathfrak{C}$ important for the exposition. We shall define these functions by introducing linear spaces K^0, K^1, which are conjugates of $\mathfrak{C}^0$, $\mathfrak{C}^1$, i.e., possess a basis $\{x^k\}$, $\{e^k\}$, for whose elements the *pairing* is defined with the base elements of $\mathfrak{C}$:

$$\langle x_k, x^j \rangle = \delta_k^j, \qquad \langle e_k, e^j \rangle = \delta_k^j, \tag{3}$$

where δ_j^k is Kronecker delta. If now

$$\alpha = \sum \alpha_k x^k, \qquad \beta = \sum \beta_k e^k \tag{4}$$

are *cochains* of dimensions 0 and 1, respectively, i.e., linear combinations, belonging to K^0, K^1, then their values

$$\alpha|x_j = \langle x_j, \alpha \rangle = \alpha_j, \qquad \beta|e_k = \langle e_k, \beta \rangle = \beta_k$$

are defined on the elements x_j, e_k belonging to $\mathfrak{C}$, i.e., **functions** over $\mathfrak{C}$ are specified. Pairing (3) is linearly extended to cochains (4) so that we have

$$\langle a, \alpha \rangle = \sum a^k \alpha_k, \qquad \langle b, \beta \rangle = \sum b^k \beta_k$$

for chain-cochain pairs. It is sometimes convenient to consider pairing to be extended to pairs of different dimensions regarding the result to be zero. In what follows, we call cochains *forms*, emphasizing their relationship with the corresponding continual objects, differential forms. The operation ∂ in $\mathfrak{C}$ induces the dual operation d in $K = K^0 \oplus K^1$:

$$\langle \partial b, \alpha \rangle = \langle b, d\alpha \rangle. \tag{5}$$

By definition, $d\beta = 0$ for any 1-form β. The operator d is an analog of the exterior differentiation operator which coincides with an ordinary derivative in a one-dimensional case.

Using (3), (5), we immediately get

$$dx^k = e^{\sigma k} - e^k, \qquad d\alpha = \sum \alpha_k dx^k = \sum (\alpha_{\tau k} - \alpha_k) e^k \tag{6}$$

or

$$d\alpha|e_k = \alpha_{\tau k} - \alpha_k. \tag{7}$$

The pairing $\langle b, \beta \rangle$ is an analog of the integral of the 1-form β with respect to some collection of intervals $b = \sum b^k e_k$, taken with the "weights" b^k (if $b^k \geq 0$ for all k, then we can suppose that these weights are measures of the corresponding intervals). If $b = \sum_1^N e_k$, then

$$\langle b, d\alpha \rangle = \alpha_{N+1} - \alpha_1 = \langle \partial b, \alpha \rangle$$

and relation (5) is an analog of the Newton–Leibniz formula (in a multidimensional case the corresponding reasoning gives the analog of the so-called general Stokes theorem (Cartan theorem). It is useful to suppose that the term $\langle \partial b, \alpha \rangle$ is also an "integral" on the 0-form over the zero-dimensional chain ∂b.

1.2. Multiplications. Let us now introduce in K a **multiplication** which is an analog of the pointwise multiplication for the 0-form (the functions of the point) and an analog of the exterior multiplication for the 1-form. In terms of the homology theory, this is the so-called Whitney multiplication [30]. Its merit is the consistency with the operation d (see relation (9) below). At the next stage we shall use it to define the analog of the **conjugation operation** (or the star operation $*$, see 3.5, Ch. 1 and 1.1, Ch. 2) and the scalar multiplication of forms. The constructions being used may seem to be rather complicated and unnatural, but it is precisely these constructions that make it possible to follow far enough the analogy with continual objects. This will become clear when we come to Ch. 3.

We shall denote the multiplication indicated above by $\smile$ and introduce it in accordance with the law

$$x^k \smile x^k = x^k, \qquad x^k \smile e^k = e^k, \qquad e^k \smile x^{\tau k} = e^k, \tag{8}$$

supposing the product to be zero in all other cases. To arbitrary forms multiplication can be extended linearly.

REMARK. Law (8) will look more natural if we write the interval e^k as a pair of its endpoints $x^k x^{\tau k}$.

PROPOSITION 1. *The relation*

$$d(\varphi \smile \psi) = d\varphi \smile \psi + \varphi \smile d\psi \tag{9}$$

is valid for the cochains $\varphi, \psi \in K$.

PROOF. It is sufficient to verify (9) for the 0-forms. In other cases both sides of the relation are zeros. Let us compare the values of the left-hand and right-hand sides of (9) on the arbitrary element e_k. If

$$\varphi \smile \psi = \sum \varphi_k \psi_k x^k,$$

then, according to (8), (6), we have

$$d(\varphi \smile \psi) = \sum \varphi_k \psi_k dx^k = \sum (\varphi_{\tau k}\psi_{\tau k} - \varphi_k \psi_k) e^k,$$

$$d\varphi \smile \psi = \sum (\varphi_{\tau k} - \varphi_k)\psi_{\tau k} e^k,$$

$$\varphi \smile d\psi = \sum \varphi_k (\psi_{\tau k} - \psi_k) e^k,$$

whence follows (9). ■

REMARK. Relation (9) preserves its form in the n-dimensional case as well, the only difference being that a factor $(-1)^p$, where p

is the degree of φ, appears before the second term of the right-hand side.

Let us now introduce the "star" operation setting

$$*x^k = e^k, \qquad *e^k = x^{\tau k}, \tag{10}$$

so that in any case

$$x^k \smile *x^k = e^k \smile *e^k = e^k. \tag{11}$$

To arbitrary forms the operation is extended linearly.

REMARK. It follows from (10) that $**x^k = x^{\tau k}$, $**e^k = e^{\tau k}$, i.e., the operation $(*)^2$ is equivalent to a shift. This is one of the main distinctive features of the formalism we introduce as compared to the continual case, where the operation $*$ is either an involution or anti-involution, i.e., $(*)^2 = \pm 1$. The possibility of "involute" definition of $*$ is discussed in Sec. 5, Ch. 3.

It immediately follows from (11) that for the φ, ψ forms of the **same** degree the relation

$$\varphi \smile *\psi = \sum \varphi_k \psi_k e^k \tag{12}$$

always holds true. Property (12) of the $\smile$ and $*$ operations serves as the basis for a "correct" (correctly imitating the continual case) definition of an inner product over the "domain" V. The sum

$$V = \sum_1^N e_k, \tag{13}$$

usually plays the part of V and the *inner product* (for forms of the same degree) is defined by the relation

$$(\varphi, \psi)_V = \langle V, \varphi \smile *\psi \rangle. \tag{14}$$

It follows from (12), (13) that

$$(\varphi, \psi)_V = \sum_1^N \varphi_k \psi_k, \tag{15}$$

and all axioms of the inner product are satisfied. It is sometimes convenient to consider the inner product to be defined for forms of different degrees too, setting it, in this case, equal to zero.

The definitions of the domain V of form (13) and the inner product (14), (15) turns the linear space of the forms over V into finite-dimensional Hilbert spaces H^0, H^1 with the bases $\{x^k\}_1^N$, $\{e^k\}_1^N$. These spaces will be of utmost importance in the exposition.

1.3. Equations and problems. Now we can pass to models of elementary differential equations and boundary value problems for them. In this case our models automatically become difference analogs of the above-mentioned objects. The introduction of a scale (normalization) and the limiting process will be discussed in the next subsection.

Having fixed the domain V defined by (13) and using the definition of the inner product, we introduce an operator δ which is a conjugate of d and the most important element of the constructions given below. Let $\alpha \in K^0$, $\varphi \in K^1$. We write a chain of relations

$$(d\alpha, \varphi)_V = \langle V, d\alpha \smile *\varphi\rangle = \langle V, d\{\alpha \smile *\varphi\}\rangle - \langle V, \alpha \smile d * \varphi\rangle =$$

$$\langle \partial V, \alpha \smile *\varphi\rangle - \langle V, \alpha \smile *\{*^{-1} d * \varphi\}\rangle. \tag{16}$$

We represent the last term of the chain as $(\alpha, \delta\varphi)_V$ by introducing the operator $\delta: K^1 \to K^0$ with the aid of the relation

$$\delta\varphi = - *^{-1} d * \varphi.$$

If we additionally suppose that the "boundary terms" $\alpha \smile *\varphi|\partial V$ vanish, then (16) gives the ordinary relation

$$(d\alpha, \varphi)_V = (\alpha, \delta\varphi)_V,$$

that connects the conjugate operators d and δ. The "pointwise" definition of δ is obviously given by the relation

$$\delta\varphi|x_k = \varphi_{\sigma k} - \varphi_k.$$

Here is the "explicit" notation of the identity given by (16):

$$\sum_1^N (\alpha_{\tau k} - \alpha_k)\varphi_k = \alpha_{\tau N}\varphi_N - \alpha_1\varphi_0 + \sum_1^N \alpha_k(\varphi_{\sigma k} - \varphi_k). \tag{17}$$

It is easy to see that (17) is an ordinary Abel transformation, namely, a discrete analog of integration by parts. Whereas (5) was an analog of the Stokes formula, (16) is an analog of the Green formula for formally conjugate differential operators d, δ (Sec. 1, Ch. 2). A remarkable property of relation (16) is that it preserves its form in a multidimensional case (Secs. 2, 3, Ch. 3), and then relations of form (17) give different generalizations of the classical Abel transformation (cf. 3.2, Ch. 3).

When using relations (16), (17) at the same time as the spaces $H^0(V)$, $H^1(V)$ introduced in 1.2, we must bear in mind the following circumstance: although inner products are taken over V of the form (13), the notation of identity (17) includes the members $\alpha_{\tau N}$, φ_0 which **are not specified** by the definition of the forms α, φ as elements of the spaces $H^0(V)$, $H^1(V)$. For relations (16), (17) to acquire the exact meaning, the values of these numbers must be defined by the requisite additional conditions.

Let us now consider the equations

$$\begin{aligned} d\alpha + \lambda * \alpha = f, \quad \delta\varphi - \lambda * \varphi = g, \\ \delta d\beta + \lambda\beta = p, \quad d\delta\psi + \lambda\psi = q, \end{aligned} \tag{18}$$

which are analogs of the differential equations

$$(\pm D_x + \lambda)u = r, \qquad (-D_x^2 + \lambda)v = s. \tag{19}$$

We suppose λ to be real. The following two circumstances must be emphasized.

1. Every one of equations (19) is associated with two analogs.
2. To write these analogs in our formalism, the operations $*$ and δ are needed in addition to the simplest "differentiation" d.

We shall consider Eqs. (18) over V of form (13) in the spaces $H^0(V)$, $H^1(V)$ assuming that $f, p \in H^1$, g, $q \in H^0$. Every equation of (18) will be associated with a certain chain of equalities. For example, for the first equation of the second row we shall have

$$2\beta_k - \beta_{\tau k} - \beta_{\sigma k} + \lambda\beta_k = p_k, \qquad k = 1, \ldots, N. \tag{20}$$

Thus, if $\beta \in H^0(V)$, then all terms in (20) will be defined only when the values of β_0, $\beta_{\tau N}$ are additionally defined. Similar "deficient" elements appear in the other equations of (18).

DEFINITION. We say that the elements $\alpha \in H^0$, $\varphi \in H^1$ are *solutions of the Cauchy problem* in V for the first pair of equations of (18) if they satisfy the corresponding chains of relations in which we additionally set

$$\alpha_{\tau N} = 0, \qquad \varphi_0 = 0. \tag{21}$$

This definition also defined the operators

$$(d + \lambda *) : \ H^0 \to H^1, \qquad (\delta + \lambda *) : \ H^1 \to H^0. \tag{22}$$

PROPOSITION 2. *For the first pair of Eqs.* (18) *the Cauchy problem is uniquely solvable for any* $f \in H^1$, $g \in H^0$. *Operators* (22) *are conjugate, i.e.,*

$$(\{d + \lambda *\}\alpha, \varphi)_V = (\alpha, \{\delta + \lambda * \varphi\})_V.$$

The first part of the statement is trivial, and the second part follows from (16), conditions (21), and the identity

$$(*\alpha, \varphi)_V = (\alpha, *\varphi)_V - \alpha_1\varphi_0 + \alpha_{\tau N}\varphi_N. \blacksquare$$

Let us consider the second part of Eqs. (18).

DEFINITION. We say that the elements $\beta \in H^0$, $\psi \in H^1$ are *solutions of the "Dirichlet problem"* in V for the second pair of Eqs. (18) if they satisfy the corresponding chains of relations in which we additionally set

$$\beta_0 = \beta_{\tau N} = 0, \qquad \psi_0 = \psi_{\tau N} = 0. \tag{23}$$

As above, this definition specifies the operators

$$(\delta d + \lambda) : \ H^0 \to H^0, \qquad (d\delta + \lambda) : \ H^1 \to H^1. \tag{24}$$

When establishing the properties of the "Dirichlet problem" and those of operators (24), it is convenient to use the identity

$$(d\alpha, d\beta)_V = \sum_1^N (\alpha_{\tau k} - \alpha_k)(\beta_{\tau k} - \beta_k) =$$

$$\alpha_{\tau N}(\beta_{\tau N} + \beta_N) - \alpha_1(\beta_1 - \beta_0) + \sum_1^N \alpha_k(2\beta_k - \beta_{\sigma k} - \beta_{\tau k}). \tag{25}$$

It results from (17) upon the substitution $\varphi_k = \beta_{\tau k} - \beta_k$. Setting $\alpha = \beta$ and taking into account (23), we obtain from (25) the relation

$$(\delta d\beta, \beta)_V = (d\beta, d\beta)_V + \beta_1^2. \tag{26}$$

At the same time, for α different from β, but with (23) taken into account, we have

$$(\delta d\beta, \alpha)_V = -\alpha_1\beta_1 + (d\beta, d\alpha)_V = (\beta, \delta d\alpha)_V. \tag{27}$$

The corresponding arguments for the second equation of the second row of (18) are completely similar. As a result we have the following statement.

PROPOSITION 3. *For $\lambda \geq 0$, $p \in H^0$, $q \in H^1$ the "Dirichlet problem" for the second pair of equations* (18) *is always uniquely solvable. Operators* (24) *are self-conjugate.*

PROOF. We shall speak of the first equation since the second part of the statement follows from (27). The first part of the statement follows from (26) and from the self-conjugacy of the corresponding operator (it is obvious that it can also be directly verified in a trivial way). ■

1.4. Norms, step functions, and approximations. Let us consider the scheme of establishing the relationship between the purely combinatorial arguments from items 1.1–1.3 and continual objects. Suppose that (a, b) is an interval of the real axis and $h = (b-a)/N > 0$. We divide the real line by the points $x_k = a + kh$, $k = 0, \pm 1, \pm 2, \ldots$. We identify the points x and the intervals $e_k = (x_k, x_{\tau k})$ with the combinatorial objects considered above. Then we associate the discrete forms $\alpha \in K^0$, $\varphi \in K^1$ with the spline functions $\alpha^h(x)$, $\varphi^h(x)$ assuming that

$$\alpha^h(x) = \alpha_k, \qquad \varphi^h(x) = \varphi_k$$

for $x \in e_k$. We consider the step functions defined in this way over (a, b) as elements of the corresponding functional Hilbert spaces $\mathbb{H}^0$, $\mathbb{H}^1$ of square-summable functions. It is obvious that

$$|\alpha^h, \mathbb{H}^0|^2 = h|\alpha, H^0|^2, \qquad |\varphi^h, \mathbb{H}^1|^2 = h|\varphi, H^1|^2,$$

where $| \, , \, |$ are norms in the corresponding Hilbert space. We define difference operators over the functions we have introduced setting

$$\Delta^h \alpha^h(x) = h^{-1}[\alpha^h(x+h) - \alpha^h(x)],$$

and do the same for φ^h. Of course, these operators are defined only in the **open** intervals e_k. In the sequel this is always implied without explicit stipulations. Let us now define the operators d^h, δ^h setting $d^h \alpha^h = \Delta^h \alpha^h$, $\delta^h \varphi^h = -\Delta^h \varphi^h$.

PROPOSITION 4. *If the relations*

$$h\alpha = \delta\varphi, \qquad h\psi = d\beta$$

are satisfied for the discrete forms α, φ, then the relations

$$\alpha^h = \delta^h \varphi^h, \qquad \psi^h = d^h \beta^h$$

are pointwise satisfied for the splines defined above.

The validity of this statement is obvious. ■

Let us now introduce the converse process, i.e., associate a continual object with a discrete one. Suppose that $f(x)$ is an integrable function of x defined over a certain collection of intervals e_k. Let us associate f with the step function f^h setting

$$f^h(x) = \frac{1}{h} \int_{e_k} f(\xi)\, d\xi \quad \text{for} \quad x \in e_k.$$

Since f defined the element of $\mathbb{H}^0$ or $\mathbb{H}^1$ (in a one-dimensional case these spaces are not traditionally distinguished), we can associate it with the corresponding object in H^0, H^1, associating the value of f^h with the point x_k or with the interval e_k. We call this procedure *discretization.*

Let us see how we can use the concepts that we introduced when constructing the approximations of solutions of continual (differential) equations. Here is a simple example, namely, the Cauchy problem for the equation $d\alpha \equiv D_x\alpha(x) = f(x)$, $x \in (a,b)$, $\alpha(b) = 0$ (the "inverse Cauchy problem").

It should be pointed out that it is often useful to use, alongside the norm $\mathbb{H}$, the analog of the W^1-norm [19, 43] defined by the relation

$$|\alpha^h, W|^2 = \int_a^b (\Delta^h \alpha^h)^2\, dx.$$

PROPOSITION 5. *Suppose that the step function f^h is the discretization of the arbitrary element $f \in H^1$. Then the Cauchy problem for the equation*

$$d^h \alpha^h = f^h \tag{28}$$

on (a,b) is uniquely solvable and the inequality

$$|\alpha^h, W| \le |f, H^1| \tag{29}$$

is satisfied for the solution of (28).

PROOF. We understand the Cauchy problem for (28) on (a,b) to be the corresponding problem in $V = \sum_1^N e_k$, i.e., the unique solvability follows from statements 4 and 2. In order to obtain estimate

(29), it is sufficient to consider the chain of relations

$$|\alpha^h, W|^2 = \int_a^b (\Delta^h \alpha^h)^2\, dx = h \sum_1^N \left(\frac{\alpha_{\tau k} - \alpha_k}{h}\right)^2 =$$

$$h \sum_1^N \left(\frac{1}{h} \int_{e_k} f(\xi)\, d\xi\right)^2 \leq \sum_1^N \int_{e_k} f^2(\xi)\, d\xi = |f, H^1|^2. \blacksquare$$

Proceeding from the splines, we shall construct "smooth" (of the class C^1) functions using the simplest averaging operator [19]. We set

$$J^h \alpha^h(x) = \frac{1}{h} \int_x^{x+h} \alpha^h(\xi)\, d\xi.$$

In a one-dimensional case the construction of approximations of the solution of the equation $d\alpha = f$ with the aid of the solution of (28) is simplified by the fact that

$$dJ^h \alpha^h(x) \equiv d^h \alpha^h(x) = f^h.$$

PROPOSITION 6. *As $h \to 0$, the family of functions $\{J^h \alpha^h\}$ converges in H^0 to the element α which has a generalized derivative and satisfies the equation $d\alpha = f$ and the condition $\alpha(b) = 0$.* $\blacksquare$

Since the solution of Eq. (28) can be easily explicitly written out in the example under consideration, it is easy to verify Proposition 6. In less trivial cases, the proof of the corresponding statement turns out to be rather laborious (cf. 3.6, Ch. 3). It should be pointed out that the elementary averaging of step functions gives the simplest **splines**.

2. Models on a Circle

It should be noted that whereas the constructions from Sec. 1 were basically connected with the elements of the homology theory, the foundation of Sec. 2 is, in essence, the theory of linear representations of finite Abelian groups [42] in its simplest (although, as in Sec. 1, somewhat unusual) form.

Suppose that S is an oriented circle of unit length divided by N points into N equal arcs. As distinct from Sec. 1, it is now convenient to have, from the very beginning, a certain "scale". We assume N to be odd and, setting $N = 2l + 1$, number the points (having chosen a certain initial point) in their sequential order by the integers $-l, -l+1, \ldots, l-1, l$, calling these integers the *coordinates* of the indicated points in the set Ω, understanding Ω as a collection of all points numbered in this way. Let f be a complex function over Ω,

$$f : \Omega \to \mathbb{C}, \qquad x \mapsto f(x), \tag{1}$$

where x is a point of Ω identified with its coordinates. The collection of functions defined in this way possesses the natural structure of a complex linear space, i.e.,

$$(\alpha f + \beta g)(x) = \alpha f(x) + \beta g(x),$$

and we shall turn it into the Hilbert space H (finite-dimensional) defining the inner product by the relation

$$(f, g) = N^{-1} \sum_x f(x) \overline{g}(x). \tag{2}$$

Then the collection of functions $\{e_x\}$, $x \in \Omega$,

$$e_x(y) = \begin{cases} \sqrt{N}, & x = y, \\ 0, & x \neq y, \end{cases} \tag{3}$$

is an orthonormal basis of H, and we can write the arbitrary element $f \in H$ as

$$\begin{gathered} f(x) = \sum_x f_x e_x, \\ f_x = (f, e_x) = N^{-1} \sum_y f(y) e_x(y) = N^{-1/2} f(x). \end{gathered} \tag{4}$$

The normalizing factor N^{-1} we have introduced in definition (2) of the inner product "damages" many of the relations that follow, making it necessary, in particular, to distinguish the value $f(x)$ of the mapping f at the point x and the coefficient f_x of the decomposition of this mapping-function into components with respect to the natural basis (3). Nevertheless, we need this normalization for the comparison (at least on the intuitive level) of discrete objects with continual ones which is of interest to us. It is precisely due to the normalization that we see that upon the indicated comparison connected with the subdivision ($N \to \infty$), the inner product (2) will be associated with integration and the element e_x (or $\sqrt{N}e_x$) will turn into the δ-function which is no longer an element of the Hilbert space. Decomposition (4) loses its direct meaning but its use in a discrete variant is very instructive.

If we introduce into H the *multiplication operator*

$$M:\ H \to H, \qquad Mf(x) = m_x f(x),$$

where m_x is a complex number, then e_x is a proper element of the operator M corresponding to the eigenvalue of m_x. In other words, supposing, as usual, that in a finite-dimensional linear space the operators are associated (for a fixed basis) with matrices, we find that in a natural basis the operator M is defined by a diagonal matrix with elements $m_{-l}, m_{-l+1}, \dots, m_l$.

We have not yet used the special structure of Ω, namely, the arrangement of the equally spaced points x on a circle. Formally, this structure reduces to the presence in Ω of the shift

$$\tau:\ \Omega \to \Omega, \qquad x \mapsto x+1, \quad x \neq l, \qquad l \mapsto -l,$$

that implies the identification of the "point" $l+1$ with the point $-l$. This shift generates, in turn, the mapping of the space of functions over Ω which is customarily defined by the law

$$\hat{\tau}f(x) = f(\tau^{-1}x).$$

The corresponding operator $\tau:\ \hat{H} \to H$ is invertible and the relation

$$(f, g) = (\hat{\tau}f, \hat{\tau}g),$$

is satisfied for it, i.e., it is *unitary*.

The central theme for considerations that refer to the model we have introduced is connected with the use of the basis H consisting of the eigenvectors of the operator $\hat{\tau}$.

PROPOSITION 1. *The functions*

$$\eta_k(x) = e^{\nu k x}, \qquad k = 1, \dots, N, \quad \nu = 2\pi i N^{-1},$$

are the eigenfunctions of the operator $\hat{\tau}$ and form an orthonormal basis of the space H.

PROOF. We have

$$\hat{\tau}\eta_k(x) = e^{\nu k(x-1)} = e^{-\nu k}\eta_k(x).$$

Next, using the properties of the geometric progression, we have

$$\sum_{x\in\Omega} e^{\nu kx} = e^{-\nu kl}\sum_{s=0}^{N-1} e^{\nu ks} = 0,$$

for the integer $k \neq 0$, whence it follows that

$$(\eta_k, \eta_j) = N^{-1}\sum_{x\in\Omega} e^{\nu(k-j)x} = \begin{cases} 1, & k = j, \\ 0, & k \neq j. \end{cases}$$

Thus the functions $\eta_k(x)$ are pairwise orthogonal, normed, and their number coincides with the dimension of the space. ■

We write the resolution of the elements of H into components with respect to the obtained basis of the eigenvectors of the operator $\hat{\tau}$ in a special form which yields an analog of the classical Fourier transform. We shall consider the numbers $1, \ldots, N$ as points, which are the coordinates of the set $\tilde{\Omega}$, equipped with the same additional structure as Ω, i.e., a finite-dimensional Hilbert space $\tilde{H}$ of complex functions with the inner product $N^{-1}\sum_k f(k)\overline{g}(k)$ is defined over $\tilde{\Omega}$ and shifts are specified which give the identification $N + 1 \equiv 1$, etc.

Then it immediately follows from Proposition 1 that the elements $\eta_x(k) = e^{\nu kx}$, $x \in \Omega$, form an orthonormal basis of the space $\tilde{H}$.

DEFINITION. We call the element $\tilde{f}(k) \in \tilde{H}$, defined by the relation

$$\tilde{f}(k) = N^{-1/2}\sum_x f(x)\eta_k(x),$$

the *Fourier transform* of the element $f(x) \in H$ and introduce the notations $\tilde{f}(k) = Ff(x)$, $F : H \to \tilde{H}$.

PROPOSITION 2. *For the operator F there exists an inverse $F^{-1} : \tilde{H} \to H$, defined by the relation*

$$\tilde{g}(x) = F^{-1}g(k) = N^{-1/2}\sum_k g(k)\overline{\eta}_x(k).$$

In order to prove this statement, it is sufficient to write the chain of relations

$$F^{-1}(Ff(x)) = N^{-1/2}\sum_k [Ff](k)\overline{\eta}_x(k) =$$

$$N^{-1}\sum_k\sum_y(f(y)e^{\nu ky})e^{-\nu kx} = N^{-1}\sum_y f(y)\sum_k e^{\nu k(y-x)} = f(x). \blacksquare$$

PROPOSITION 3. *The operator F is unitary.*
Indeed,

$$(Fg, Ff) = N^{-2}\sum_k\left(\sum_x f(x)\eta_k(x)\sum_y\overline{g(y)\eta_k(y)}\right) =$$

$$N^{-1}\sum_x f(x)\overline{g}(x). \blacksquare$$

In many problems that refer to the interconnection of the spaces H and $\tilde{H}$, an essential part is played by the convolution operation defined in H by the relation

$$f * g = N^{-1}\sum_y f(x-y)g(y).$$

This is due to the fact that in the spaces $H, \tilde{H}$ the operation of **pointwise** multiplication $g(x)f(x)$ of the elements of H and $\tilde{H}$, that turns $H, \tilde{H}$ into **algebra**, can always be introduced alongside the multiplication operator M considered above.

REMARK. It should be borne in mind that in a continual case, for the pointwise multiplication of two elements of a functional Hilbert space to be defined, at least one of them must meet certain special additional requirements.

PROPOSITION 4. *The relation*

$$\widetilde{f * g} = N^{-1/2}\tilde{g}\cdot\tilde{f} \quad or \quad \widetilde{f * g}(k) = N^{-1/2}\tilde{f}(k)\tilde{g}(k),$$

holds true, i.e., the Fourier transformation associates the convolution operation with that of pointwise multiplication.

PROOF. We have

$$\widetilde{f * g}(k) = N^{-3/2}\sum_x\left(\sum_y f(x-y)g(y)\right)e^{\nu kx} =$$

$$N^{-3/2}\sum_x\sum_y f(x-y)e^{\nu k(x-y)}g(y)e^{\nu ky} =$$

$$N^{-3/2}\sum_y g(y)e^{\nu ky}\sum_x f(x-y)e^{\nu k(x-y)} = N^{-1/2}\tilde{g}(k)\tilde{f}(k). \blacksquare$$

As a result of a similar verification (or directly from Proposition 4) we obtain the following statement.

PROPOSITION 5. *The relation*

$$\sqrt{N}\widetilde{f \cdot g} = \tilde{f} * \tilde{g}$$

holds true.

PROOF.

$$\tilde{f} * \tilde{g} = N^{-1} \sum_p \tilde{f}(k-p)\tilde{g}(p) = N^{-1} \sum_p \sum_x f(x) e^{\nu x(k-p)} \sum_y g(y) e^{\nu y p} =$$

$$N^{-1} \sum_x f(x) \sum_y g(y) \sum_p e^{\nu p(y-x)} e^{\nu k x} = \sum_x f(x) g(x) e^{\nu k x}. \blacksquare$$

PROPOSITION 6. *The element* $\sqrt{N} e_0$ *plays the part of unity relative to the convolution operation.*

PROOF.

$$f(x) * e_0 = N^{-1} \sum_y f(x-y) e_0(y) = N^{-1/2} f(x). \blacksquare$$

In our formalism this corresponds to the fact that

$$\tilde{e}_x(k) = N^{-1/2} N^{1/2} e^{\nu k x}, \qquad \tilde{e}_0 = 1,$$

$$\tilde{f} \cdot \tilde{e}_0 = \sqrt{N} \widetilde{f * e_0} = \widetilde{f * \sqrt{N} e_0} = \tilde{f}.$$

As would be expected, the Fourier transformation that uses the expansion of the shift operator in eigenfunctions proves to be a natural instrument for investigating large classes of difference equations, the simplest of which is the equation (system)

$$u(x+1) - u(x) - \lambda u(x) = f(x), \qquad x \in \Omega, \tag{5}$$

with the unknown function $u(x)$. Setting

$$u(x) = N^{-1/2} \sum_k \tilde{u}(k) e^{\nu k x}$$

(we can obviously interchange the parts played by the operators F and F^{-1}) and writing the similar representation for $f(x)$, we obtain

$$\tilde{u}(k)(e^{\nu k} - 1 - \alpha) = \tilde{f}(k), \tag{6}$$

or, on the assumption that

$$e^{\nu k} - 1 - \lambda \neq 0, \qquad k = 1, \ldots, N,$$

we have the representation of the solution

$$u(x) = N^{-1/2} \sum_k \tilde{f}(k)(e^{\nu k} - 1 - \lambda)^{-1} e^{\nu k x},$$

where $f = F^{-1} f(x)$ is a known function.

Writing (5) in operator form

$$Lu - \lambda u = f,$$

we see that the values of λ

$$\lambda_k = e^{\nu k} - 1, \qquad k = 1, \ldots, N,$$

are the eigenvalues of the operator L. For

$$\lambda = e^{\nu k_0} - 1$$

the homogeneous equation (5) has the nontrivial solution

$$u_0 = e^{\nu k_0 x}.$$

It follows from (6) that in a certain sense the Fourier transformation associates the difference operator with the operation of "multiplication by an independent variable".

On the other hand, statements 4, 5 show that the transition from u to $\tilde{u}$ is an obvious means of solving the equation in convolutions

$$u * g = f.$$

We have

$$\widetilde{u * g} = N^{-1/2} \tilde{u}\tilde{g} = \tilde{f},$$

and, if $\tilde{u} = Fu$, then $u = F^{-1}(\sqrt{N}\tilde{f}/\tilde{g})$.

At the same time we can note the difficulty encountered in the attempt to use this approach for solving equations of the form (5) which contain the "variable coefficient" $\lambda = \lambda(x)$. The equation resulting from the Fourier transformation includes the convolution $\tilde{\lambda} * \tilde{u}$.

Chapter 1

Formal Structures

0. Introductory Remarks

By no means can this chapter replace textbooks and monographs that contain a detailed study of the structures considered below. I can only hope that the exposition I present can serve as a summary of necessary data, rather formal in places, but, at the same time, containing a number of informal remarks that emphasize the specific features of the chosen point of view. Additional explanations are given in the "0" items prefacing every section.

In the definitions I repeatedly use the terms a "set" and a "correspondence" (or their synonyms such as a "collection", a "tuple" for a set, and a "function" and a "mapping" for a correspondence). Proceeding from the "naive" viewpoint, I consider these terms to be primary, intuitively clear, that are not subject to formalization or any further logical analysis at this level. The remarks concerning the set-theoretic symbols which are used are given in 1.0.

Two classical sets, the sets $\mathbb{R}$ and $\mathbb{C}$ of real and complex numbers, play an exceptionally significant part in constructing one or another type of abstract objects or in defining some additional structure on it (that makes the object more informative). I suppose that their properties are well known.

The references to papers that I consider to be the most suitable are given inside the sections.

1. Topology and Metrics

1.0. Preliminary notes. Since the main subject matter of this book is the classical (multidimensional) analysis using functions defined on subsets of the Euclidean space (or the Riemannian manifold which is its direct generalization), the major part of the discussion that follows deals with structures that are considerably richer than the one defined only by introduction of the topology. Nevertheless, such an object as a topological space and the corresponding general definition of continuity (continuous mapping) are very convenient initial concepts that are repeatedly used in the sequel.

For example, an attempt to define a differentiable manifold without using the term "Hausdorff topological space" leads to a cumbersome construction.

At the same time, the following must be pointed out in addition to what was said in the introduction. We suppose that as applied to the sets X and Y, such terms as an inclusion ($X \subset Y$), a union ($X \cup Y$), an intersection ($X \cap Y$), taking the complement or the difference ($X \setminus Y$) have no need of any comments. The empty set is denoted by $\varnothing$.

Unless otherwise specified, the mapping $f: X \to Y$ is always considered to be **unique**. Thus if $Y' \subset Y$ is a subset and $f^{-1}(Y') \subset X$ is its inverse image under the mapping f, then in this notation f^{-1} is not, strictly speaking, a mapping. If we are given a function $u(y)$ over Y and a mapping $f: X \to Y$, then $f^*u(x)$ is the corresponding function over X.

The frequently used term the *direct product* $X \times Y$ of the sets X and Y means a set whose elements are ordered pairs (x, y), $x \in X$, $y \in Y$.

We can consider [24] to be a standard textbook containing a detailed discussion of preliminary concepts.

1.1. Topological space. A *topological space* is a set X in which a family of subsets $\{O_\alpha\}$ is chosen. These subsets are called *open* and satisfy the following requirements.

0–1. The union of any collection of open sets is open.

0–2. The intersection of any finite number of open sets is open.

0–3. The set X and the empty set $\varnothing$ are open.

The family $\{O_\alpha\}$ is known as a *topology*. The topology $\{O'_\beta\}$ is *stronger* than the topology $\{O_\alpha\}$ if every set that is open in the topology $\{O_\alpha\}$ is open in the topology $\{O'_\beta\}$. The strongest topology in X is a *discrete* topology (every set is open) and the weakest topology is a *trivial* one (only X and $\varnothing$ are open). Two topologies are *equivalent* if each of them is stronger than the other. Note that two different topologies in X must not necessarily be *comparable*.

The subset S of a topological space is *closed* if $X \setminus S$ is open. To emphasize the geometric meaning of formal definitions, the elements of a topological space are usually called **points**. The open set O containing a point x ($x \in O$) is called a *neighborhood* of this point. The point $x \in X$ is a *limit point* for the subset $S \subset X$ if any neighborhood of the point x contains at least one point $y \in S$ different from x.

PROPOSITION 1. *A closed set contains all of its limit points.* ■

A minimal closed set that contains the given set $S \subset X$ is the *closure* of S (usually denoted by $\mathrm{cl}\, S$ or $\overline{S}$).

The sequence of points $\{x_k\}_1^\infty$ *converges* to the point $x \in X$ if, for any neighborhood U_x of the point x, there exists a number N such that $x_k \in U_x$ for $k > N$.

O–H. The topological space X is a *Hausdorff* space if any two of its different points have nonintersecting neighborhoods.

Axiom O–H ensures the **uniqueness** of the limit of a convergent sequence.

Let X, Y be topological spaces. The mapping $f: X \to Y$ is *continuous* if, for any open set $U \subset Y$, its inverse image $f^{-1}(U) \subset X$ is an open set. The mapping $f: X \to Y$ is homeomorphic if it is one-to-one, maps X onto the whole space Y, and both mappings f and f^{-1} are continuous. Spaces connected by a homeomorphism are *homeomorphic*. A homeomorphism is an **equivalence relation** and the whole totality of topological spaces can be divided into classes of topologically equivalent (homeomorphic) spaces. The properties of a topological space that are preserved under an arbitrary homeomorphism (that are inherent in every element of the corresponding class) are said to be *topologically invariant*.

PROPOSITION 2. *The interval* $(0,1)$ *and the entire real axis are topologically equivalent.* ■

The system $\{U_\sigma\}$ of open sets forms an *open cover* for the set

$S \subset X$ if, for any $x \in S$, there exists at least one set U_σ such that $x \in U_\sigma$.

The set $S \subset X$ is *compact* if we can choose a finite cover from any open cover of S. A topological space is *locally compact* if every one of its points has a neigborhood whose closure is compact.

The subsets A, B of the topological space X are *disjoint* if $A \cap \overline{B} = \varnothing$, $\overline{A} \cap B = \varnothing$. The subset $Y \subset X$ is *connected* if it is not the union of two disjoint subsets.

1.2. Metric space. Despite the logical orderliness of the definitions given in 1.1, an analyst feels more confident in situations in which a topology is defined by the specification of a **metric**.

A *metric space* is a set M in which a real nonnegative function $\rho(x, y)$ of the pair of elements $x, y \in M$ is defined which satisfies the following requirements:

1. $\rho(x, y) = \rho(y, x)$.
2. $\rho(x, y) + \rho(y, z) \leq \rho(x, z)$.
3. $\rho(x, y) = 0$ if and only if $x = y$.

The function ρ is a *metric* and its value on the pair of elements x, y is the *distance* between these elements.

The specification of a metric defines the corresponding topology in a natural way. Indeed, we say that the ε-*neighborhood* of the point $x \in M$ is the set of all $y \in M$ for which $\rho(x, y) < \varepsilon$. Then the set $U \subset M$ is *open* if, for any $x \in U$, there exists $\varepsilon = \varepsilon(x, U)$ such that the ε-neighborhood x belongs to U.

Two metrics in M are *equivalent* if they generate equivalent topologies.

A metric space is automatically a Hausdorff space.

A metric space differs from an arbitrary topological space by the possibility of introducing some additional concepts. Thus the sequence $\{x_k\}_1^\infty$ of elements of M is known as a *Cauchy sequence* (or a *fundamental sequence*) if, for any $\varepsilon > 0$, there exists a number $N = N(\varepsilon)$ such that $\rho(x_k, x_j) < \varepsilon$ for any choice of $k, j > N$.

The metric space M is *complete* if, for any fundamental sequence, there exists in it a limit that belongs to M.

It is interesting to note that the property of a sequence to be fundamental is not topologically invariant. For instance, the homeomorphic mapping $f : x \to 1/x$ of the space of positive real numbers (with a natural topology) maps the sequence $\{1/n\}_1^\infty$ into the sequence $\{n\}_1^\infty$. The concept of a **bounded set** which has an obvious natural sense in a metric space is not topologically invariant either.

In the sequel we shall encounter a large number of different means of defining a metric on sets which possess one or another additional structure.

2. Groups and Complexes

2.0. Preliminary remarks. Although the section begins with a standard general definition of a group, in the subsequent discussion we use, as a rule, a group with a commutative group operation (commutative or Abelian) as one of the most important "intermediate" structures entering into the principal definitions. The only exceptions are certain constructions from Ch. 4 and separate remarks.

The Abelian group enters into the definition of a linear space (Sec. 3) and in this section it is used for defining a complex, namely, an algebraic equivalent of a geometric formation on which functions that are of interest to an analyst are defined. It should be noted that complexes which are models of domains of the Euclidean space play the main part in Ch. 3, are very special, and are not directly connected with the examples given below. However, the construction of these special complexes is based on the elementary homological formalism presented here, and therefore it is useful to consider the examples illustrating it.

Returning to the remarks made in the introduction to this chapter, I want to point out that although the definition of a group that we use is standard, it is not completely formalized. Speaking of a set, we imply that the concept of an identity satisfying the ordinary requirements is defined for its elements and consider the meaning of such terms of an "operation", "comparison", etc. to be obvious.

In a section containing the definition of a group it would be natural to say some words about a group **representation**, which is one of the most important concepts, but from formal considerations these remarks are given in Sec. 3.

I would recommend [26, 29, 52] (groups), [21, 30, 31] (complexes, homology) as standard textbooks that contain the material considered in this section.

2.1. Definitions. A *group* is a nonempty set G of elements $u, v, w, \ldots$, in which the binary **operation** (multiplication) is defined that associates every pair u, v of the elements of G with a single element $w : uv = w$. The operation is subject to the following additional requirements: (a) the operation is associative, i.e., $u(vw) = (uv)w$; (b) there exists a right-hand neutral element e (the right-hand unit element) such that $ue = u$ for any $u \in G$; and (c) for any $u \in G$ there exists a right-hand inverse element u^{-1} such that $uu^{-1} = e$.

PROPOSITION 1. *The right-hand unit element is simultaneously a left-hand unit element and the converse right-hand unit element is a left-hand converse unit element.* ■

The mapping T of the group G_1 into the group G_2 which is consistent with the group operation $T(uv) = TuTv$ is known as a *homomorphism.* The homomorphism that defines a one-to-one correspondence of the elements of two groups is an *isomorphism.* It is not improbable that in this case $G_2 = G_1$ (*automorphism*).

A group is *Abelian* if the group operation is additionally subjected to the requirement of commutativity, i.e., $uv = vu$. In this event, the symbol "+", i.e., $u + v$ is used to denote the operation, the symbol 0: $u + 0 = u$ is used for the neutral element, and the notation $-u$: $u + (-u) \equiv u - u = 0$ is used for the inverse of u.

REMARK. The so-called topological or continuous groups originating as a result of the introduction of a topology in a group, or of a group operation, in a topological space is one of the most important objects of investigation in contemporary mathematics. In this case the requirement of **continuity** of a group operation, i.e., the requirement of the concordance of structures, is obligatory.

Proceeding from the concept of an Abelian group, we introduce the concept of a **complex**. Let $\mathfrak{C}^k$, $k = 0, \ldots, n$ be a collection of Abelian groups. In the direct product $\mathfrak{C}^0 \times \ldots \times \mathfrak{C}^n$ the operation of addition is defined in a natural way, and in this case it is customary to speak of the direct sum

$$\mathfrak{C} = \bigoplus_0^n \mathfrak{C}^k.$$

The direct sum $\mathfrak{C}$ in which the *boundary homomorphism*

$$\partial : \ \mathfrak{C}^k \to \mathfrak{C}^{k-1}, \qquad k = 1, \ldots, n, \qquad \partial \mathfrak{C}^0 = 0,$$

is defined which is subject to the additional requirement

$$\partial^2 \equiv \partial\partial = 0, \tag{1}$$

is known as an *n-dimensional complex of Abelian groups.*

The element $u \in \mathfrak{C}^k$ is said to be of the *dimension k.* Property (1) is equivalent to the existence of the inclusion

$$\operatorname{Im}\partial \subset \operatorname{Ker}\partial,$$

where $\operatorname{Im}\partial$ is the **image** and $\operatorname{Ker}\partial$ is the **kernel** (the inverse image of zero) of the homomorphism ∂. The elements $\mathfrak{C}^k$ are *chains*, the chains belonging to $\operatorname{Ker}\partial$ are *cycles*, and those belonging to $\operatorname{Im}\partial$ are *boundaries* (of the corresponding dimension). By virtue of (1), every boundary is a cycle, but the converse statement is not always true. If ∂_k is the restriction of ∂ to $\mathfrak{C}^k$, then the **quotient group**

$$\mathcal{H}^k(\mathfrak{C}) = \operatorname{Ker}\partial_k / \operatorname{Im}\partial_{k+1}$$

is called a k-dimensional *group of homology* of the complex $\mathfrak{C}$. The transition to a quotient group means that the elements of $\mathcal{H}^k$ are the classes $[c]$ of the cycles $c + \partial u$, i.e., the cycles c_1 and c_2 are identified if $c_1 - c_2 \in \operatorname{Im}\partial$. Two cycles of the same class are *homologous* and the notation $c_1 \sim c_2$ is used to denote a homology relation.

It can be immediately verified that $\mathcal{H}^k$, in turn, possesses a natural group structure. A group of homologies of the complex $\mathfrak{C}$ is defined as the direct sum of the groups $\mathcal{H}^k$, i.e.,

$$\mathcal{H}(\mathfrak{C}) = \bigoplus_0^n \mathcal{H}^k(\mathfrak{C}).$$

In the discussions that follow in this section we shall deal with Abelian groups defined by a *system of generators* $\{u_k\}$. In this case every element of the group can be represented as the sum $\sum a_k u_k$, where a_k are arbitrary integers. If the generator u is unique and the relation $au = bu$ implies that $a = b$, then the group is said to be an *infinite cyclic* group and denoted by $\mathbb{Z}_\infty(u)$. If the relation $au = bu$ is also valid for certain $a \neq b$, then there exists $p \neq 0$ such that $pu = 0$. Among all p of this kind there is a **least** p_0, and the corresponding group is said to be a *cyclic group of order* p_0 (denoted by $\mathbb{Z}_{p_0}(u)$).

The simplest example is a group $\mathbb{Z}_2$ of two elements 0 and u in which $u + u = 0$.

It is useful to note that the following theorem is valid.

THEOREM. *Every Abelian group with a finite system of generators can be represented as the direct sum of cyclic groups.* ■

In this case the assumption concerning the finiteness of the number of generators is essential.

2.2. The simplest examples of complexes. Let us elucidate the geometric meaning of the concepts that have been introduced.

2a. We take a circle S^1 divided by the points x_1, x_2 into two arcs, which are semicircles e_1, e_2. We define the mapping $\tilde{\partial}$: $\tilde{\partial} e_1 = x_1 - x_2$, $\tilde{\partial} e_2 = x_2 - x_1$, which associates an arc with its "boundary elements". Note that at the same time we defined a certain **orientation** on S^1, namely, the point x_1 is the endpoint of the arc e_1 and the point x_2 is the endpoint of the arc e_2. In addition we set $\tilde{\partial} x_1 = \tilde{\partial} x_2 = 0$.

REMARK. We stipulate that the concept of **orientation**, which will play a significant part in further constructions, must be given a formal definition which will be given at an appropriate moment. For the time being, we shall use sufficiently clear intuitive reasoning that was used above.

We define the complex $\mathfrak{C}(S^1)$ by setting

$$\mathfrak{C}(S^1) = \mathfrak{C}^0 \oplus \mathfrak{C}^1,$$

where $\mathfrak{C}^0, \mathfrak{C}^1$ are Abelian groups generated by the elements x_1, x_2, e_1, e_2, respectively. This means that

$$\mathfrak{C}^0 = \mathbb{Z}_\infty(x_1) \oplus \mathbb{Z}_\infty(x_2)$$

and a similar representation is valid for $\mathfrak{C}^1$.

We define the **boundary homomorphism** ∂ in $\mathfrak{C}(S^1)$ extending linearly the mapping $\tilde{\partial}$ we have introduced to chains of the form

$$a_1 x_1 + a_2 x_2, \qquad b_1 e_1 + b_2 e_2 \tag{2}$$

with arbitrary integral coefficients. Requirement (1) is trivially met. This completes the definition of the corresponding complex.

It is now easy to see that all zero-dimensional chains and one-dimensional chains of the form $b(e_1 + e_2)$ are cycles in $\mathfrak{C}(S^1)$ and zero-dimensional chains of the form $a(x_1 - x_2)$ are boundaries. Consequently, $x_1 \sim x_2$ (since $x_1 - x_2 = \partial e_1$), i.e., any zero-dimensional chain $a_1 x_1 + a_2 x_2$ can be written as $(a_1 + a_2)x_1$ and the class $[x_1]$

can be taken as the generator $\mathcal{H}^0(\mathfrak{C})$. For $\mathcal{H}^1(\mathfrak{C})$ the generator is $[e_1 + e_2]$. The final result is

$$\mathcal{H}(\mathfrak{C}(S^1)) = \mathbb{Z}_\infty([x_1]) \oplus \mathbb{Z}_\infty([e_1 + e_2]).$$

It is customary to formulate the result obtained as the following statement. "A circle has two fundamental cycles which are not homologous to zero, one of dimension 0 and the other of dimension 1."

The following circumstance is a remarkable property of the described construction: if we take some other **cell partitioning** of the circle S^1, dividing it by the points $x_1, \ldots, x_n$ into the arcs $e_1, \ldots, e_n$, set $\partial e_k = x_{k+1} - x_k$, $k = 1, \ldots, n-1$, $\tilde{\partial} x_n = x_n - x_1$, $\tilde{\partial} e_k = 0$, $k = 1, \ldots, n$, and repeat the given arguments for the definition of the complex $\mathfrak{C}'(S^1)$, we obtain

$$\mathcal{H}(\mathfrak{C}'(S^1)) = \mathbb{Z}_\infty([x_1]) \oplus \mathbb{Z}_\infty([e_1 + \ldots + e_n]),$$

i.e., the homology groups of the complexes $\mathfrak{C}(S^1)$, $\mathfrak{C}'(S^1)$ are obviously **isomorphic**. Consequently, the groups $\mathcal{H}^k$ characterize certain **interior** properties of an object, independent of "accidental" properties of the construction. We shall discuss the corresponding general result concerning geometric objects and the respective complexes at the end of the section. Here are some more examples of an illustrative nature.

2b. Let S^1 be a circle considered together with the simplest division given in 2a. We shall define the *disk* $\mathcal{D}$ resulting from the addition to S^1 of the "middle", i.e., the element v of dimension 2 with boundary $\tilde{\partial} v = e_1 + e_2$. The complex

$$\mathfrak{C}(\mathcal{D}) = \mathfrak{C}^0 \oplus \mathfrak{C}^1 \oplus \mathfrak{C}^2 \tag{3}$$

is characterized by the addition of the elements cv to sums (2), where c is an integer. We repeat the reasoning used for defining $\mathcal{H}(\mathfrak{C}(S^1))$ and obtain

$$\mathcal{H}(\mathfrak{C}(\mathcal{D})) = \mathbb{Z}_\infty([x_1]). \tag{4}$$

Indeed, the element v does not generate any cycles, and the cycle $e_1 + e_2$ is now homologous to zero (is the boundary of v).

REMARK. Notation (4) is not satisfactory since it lacks the indication of the dimension of the original complex $\mathfrak{C}(\mathcal{D})$ which is an important characteristic of an object. This deficiency must be compensated for either by the use of (3) or by the explicit indication of $\dim \mathfrak{C} = 2$, or else by including the zero terms into (4).

2c. We define the cylinder Q as the direct product $S^1 \times I$ of the circle by the interval I consisting of a one-dimensional element ε and the endpoints y_1, y_2. We introduce four generators $x_k \times y_j$ in $\mathfrak{C}^0(Q)$, six generators $x_k \times \varepsilon$, $e_k \times y_j$ in $\mathfrak{C}^1(Q)$, and two generators $e_k \times \varepsilon$ in $\mathfrak{C}^2(Q)$, $k, j = 1, 2$. Setting $\tilde{\partial}\varepsilon = y_2 - y_1$ and defining $\tilde{\partial}$ on the elements S^1 as we did before, we must define $\tilde{\partial}$ on the elements of Q supposing that the mapping $\tilde{\partial}$ annuls the elements of $\mathfrak{C}^0$ and that

$$\tilde{\partial}(x_k \times \varepsilon) = x_k \times \tilde{\partial}\varepsilon, \qquad \tilde{\partial}(e_k \times y_j) = \tilde{\partial}e_k \times y_j,$$
$$\tilde{\partial}(e_k \times \varepsilon) = \tilde{\partial}e_k \times \varepsilon - e_k \times \tilde{\partial}\varepsilon$$

on $\mathfrak{C}^1, \mathfrak{C}^2$.

Reasoning as before, we find that

$$\mathcal{H}(\mathfrak{C}(Q)) = \mathbb{Z}_\infty([x_1 \times y_1]) \oplus \mathbb{Z}_\infty([e_1 \times y_1 + e_2 \times y_1]),$$

i.e., the homology group Q is **isomorphic** to the homology group S^1. From the formal point of view the complexes $\mathfrak{C}(S^1)$, $\mathfrak{C}(Q)$ differ only in the dimension.

2d. If we carry out the corresponding constructions for a torus $T = S^1 \times S^1$ supposing that the second circle is associated with a complex generated by the points y_1, y_2 and the arcs $\varepsilon_1, \varepsilon_2$, and using the notations similar to those introduced for Q, we obtain

$$\mathcal{H}(\mathfrak{C}(T)) = \mathbb{Z}_\infty([x_1 \times y_1]) \oplus \mathbb{Z}_\infty([(e_1 + e_2) \times y_1]) \oplus$$
$$\mathbb{Z}_\infty([x_1 \times (\varepsilon_1 + \varepsilon_2)]) \oplus \mathbb{Z}_\infty([(e_1 + e_2) \times (\varepsilon_1 + \varepsilon_2)]).$$

2e. We define the sphere S^2 by adding to S^1 two hemispheres v_1, v_2, and the map $\tilde{\partial}v_1 = e_1 + e_2 = -\tilde{\partial}v_2$. Noting that the sum $v_1 + v_2$ in $\mathfrak{C}(S^2)$ is a cycle (whereas $e_1 + e_2$ is a cycle homological to 0) and presenting the appropriate arguments, we obtain

$$\mathcal{H}(\mathfrak{C}(S^2)) = \mathbb{Z}_\infty([x_1]) \oplus \mathbb{Z}_\infty([v_1 + v_2]).$$

In all the preceding examples we dealt with infinite cyclic groups, and now we shall consider the simplest "geometric object" for which $\mathcal{H}^2 = \mathbb{Z}_2$.

2f. The *real projective plane* P^2, regarded as a topological space, can be obtained from the sphere S^2 by means of the identification of every point with the diametrically opposite point. Turning to our algebraic model for S^2, we see that this operation corresponds to the introduction of relations

$$x_1 = x_2, \qquad e_1 = e_2, \qquad v_1 = -v_2,$$

i.e., the groups $\mathfrak{C}^0, \mathfrak{C}^1, \mathfrak{C}^2$ now have only one generator each: x, e, and v. In this case

$$\tilde{\partial} x = \tilde{\partial} e = 0, \qquad \tilde{\partial} v = 2e, \tag{5}$$

so that e is a cycle nonhomologous to zero whereas $2e \sim 0$. The result that follows from our reasoning and from (5) can be expressed by the relation

$$\mathcal{H}(\mathfrak{C}(P^2)) = \mathbb{Z}_\infty([x]) \oplus \mathbb{Z}_2([e]). \tag{6}$$

Here is the standard technique of formulating the results obtained. The number of infinite cyclic groups induced by the generators of the requisite dimensions and entering into $\mathcal{H}(\mathfrak{C})$ is known as *Betti number*. Denoting these numbers by B_k (the subscript k indicates the dimension), we see that in all these examples $B_0 = 1$. This fact corresponds to the **connectivity** of geometric figures which we have used in our constructions. As concerns the other dimensions, we shall have $B_1 = 1$ for S^1; $B_1 = B_2 = 0$ for D; $B_1 = 1$, $B_2 = 0$ for Q; $B_1 = 2$, $B_2 = 1$ for T; $B_1 = 0$, $B_2 = 1$ for S^2; and $B_1 = B_2 = 0$ for P^2.

In order to show that (6) includes the term $\mathbb{Z}_2$, we must add a new element to the given description. If it turns out that, alongside infinite cyclic groups, $\mathcal{H}(\mathfrak{C})$ also includes finite cyclic groups, then their orders are called *torsion coefficients* (of the corresponding dimensions). In our examples all torsion coefficients, except for P^2, are zeros. For P^2 the torsion coefficient of dimension 2 is equal to 2.

It is important to point out the following. We could have defined Betti numbers by using arbitrary rational or real numbers as the coefficients a_k, b_k in (2). However, upon this extension of the class of coefficients, the torsion coefficients would automatically become equal to zero in all cases. Indeed, turning to (5), we see that we can then set $e = \tilde{\partial}(v/2)$, i.e., the cycle e proves to be homologous to zero.

In a number of cases this situation inevitably arises when we use the analytic apparatus for defining homology characteristics (Ch. 2). In such cases we must sacrifice torsion. The only consolation is that in the cases we are interested in it is necessarily zero.

2.3. Cohomology. Suppose that $\mathfrak{C}^r$ is a group of r-dimensional chains of the n-dimensional complex $\mathfrak{C}$ which is an Abelian group with the system of generators $\{u_s\}$. Suppose, furthermore, that $\mathbb{R}$ is the set of real numbers regarded as an Abelian group relative to addition, and K_r is the collection of all homomorphisms $\mathfrak{C}^r \to \mathbb{R}$ which is written as the relation

$$K_r = \mathrm{Hom}\,(\mathfrak{C}^r, \mathbb{R})$$

(for an analyst K_r is the set of all real "linear" functionals over $\mathfrak{C}^r$). If $v \in K_r$, then the value of the corresponding homomorphism on the chain $u \in \mathfrak{C}^r$ is denoted by $\langle u, v\rangle$. Let $u^p \in K_r$ be a homomorphism for which $\langle u_s, u^p\rangle = \delta_s^p$ (Kronecker delta). Then the arbitrary element $\omega \in K_r$ can be represented as the sum

$$\omega = \sum_p \omega_p u^p$$

and is known as a *cochain with real coefficients* of dimension r.

The set K_r has a natural structure of an Abelian group, and we can define the *coboundary* homomorphism $d_r: \; K_r \to K_{r+1}$, generated by the homomorphism ∂ according to the law

$$\langle \partial u, v\rangle = \langle u, dv\rangle$$

and possessing the obvious property

$$dd = 0.$$

By definition, d: $K_n \to 0$. The direct sum $K = \oplus_0^n K_r$ in conjunction with the coboundary homomorphism d is called an n-dimensional *complex of cochains* with real coefficients over the complex $\mathfrak{C}$.

REMARK. Generally speaking, in the definition of K_r we can use the arbitrary group G: $K_r = \mathrm{Hom}\,(\mathfrak{C}^r, G)$ rather than $\mathbb{R}$. Then we speak of a complex K "with coefficients from G".

The *homology groups* $\mathcal{H}_k(K)$ are defined by complete analogy with the homology groups for $\mathfrak{C}$, i.e.,

$$\mathcal{H}_k = \operatorname{Ker} d_{k+1} / \operatorname{Im} d_k.$$

The application of the scheme described above to the examples considered in 2.2 shows that the corresponding groups are completely defined by homology groups and, consequently, their structure also reflects the "geometric" properties of the original complex. At the same time, the group $\mathcal{H}(K)$ does not give additional information. Moreover, in Example 2f the information on torsion is lost, as would be expected, because of the use of real coefficients.

In what follows, in Ch. 3, we will have to deal with groups K_r which are linear spaces and which, in addition, we shall turn into finite-dimensional Hilbert spaces. In algebraic topology the complex $K(\mathfrak{C})$ proves to be the carrier of important additional information due to the possibility of introducing the topologically invariant **multiplication operation** in $\mathcal{H}(K)$. We shall also use this operation (cf. the operation $\smile$ in 1.2, Ch. 0) in the special situation of interest to us.

2.4. Tensor products. The cylinder and torus considered in 2.2 were defined as the direct products $S \times I$, $S \times S$, i.e., as the product of a circle by a line segment and the product of a pair of circles. The described technique of constructing the corresponding complexes served as an example of the use of the so-called **tensor multiplication**:

$$\mathfrak{C}(Q) = \mathfrak{C}(S) \oplus \mathfrak{C}(I), \qquad \mathfrak{C}(T) = \mathfrak{C}(S) \oplus \mathfrak{C}(S).$$

A similar relationship between the direct multiplication of geometric objects and the tensor multiplication of the corresponding complexes is preserved in the general case. Here are the necessary definitions.

The *tensor product* $G_1 \otimes G_2$ *of the Abelian groups* G_1 *and* G_2, defined by the tuples of generators $\{u_k\}$, $\{v_j\}$, respectively, is a group with generators $u_k \otimes v_j$, the following relations being postulated:

$$\begin{aligned} (u_k + u_l) \otimes v_m &= u_k \otimes v_m + u_l \otimes v_m, \\ u_k \otimes (v_m + v_n) &= u_k \otimes v_m + u_k \otimes v_n, \end{aligned} \tag{7}$$

It follows that

$$a(u_k \otimes v_m) = au_k \otimes v_m = u_k \otimes av_m \tag{8}$$

for any integer a and, in particular, $0 \otimes v_k = u_m \otimes 0 = 0$. It should be emphasized that not every element of a tensor product is *prime*, i.e., representable in the form $u \otimes v$, $u_1 \in G$, $v \in G_2$.

There exists a natural isomorphism between $G_1 \otimes G_2$ and $G_2 \otimes G_1$ and between $G_1 \otimes (G_2 \otimes G_3)$ and $(G_1 \otimes G_2) \otimes G_3$, but the corresponding identifications are not obligatory. We shall avoid this identification in the first case and use it in the second (associative) case.

The *tensor product of the complexes*

$$\mathfrak{C}_{(1)} = \overset{n}{\underset{0}{\oplus}} \mathfrak{C}^k_{(1)}, \qquad \mathfrak{C}_{(2)} = \overset{m}{\underset{0}{\oplus}} \mathfrak{C}^j_{(2)}$$

is the complex

$$\mathfrak{C} = \overset{n}{\underset{0}{\oplus}} +m\, \mathfrak{C}^k, \tag{9}$$

where

$$\mathfrak{C}^k = \underset{p+q=k}{\oplus} (\mathfrak{C}^p_{(1)} \otimes \mathfrak{C}^q_{(2)}) \tag{10}$$

and the tensor products of the corresponding Abelian groups are used in the definition of $\mathfrak{C}^k$. In this event, if c^p, c^q are chains of the indicated dimension, belonging to the complexes being multiplied, then

$$\partial(c^p \otimes c^q) = \partial c^p \otimes c^q + (-1)^p c^p \otimes \partial c^q. \tag{11}$$

Since

$$\partial\partial(c^p \otimes c^q) =$$

$$\partial\partial c^p \otimes c^q + (-1)^{p-1} \partial c^p \otimes \partial c^q + (-1)^p \partial c^p \otimes \partial c^q + (-1)^p c^p \otimes \partial\partial c^q = 0,$$

relation (1) is satisfied and the definition of the complex $\mathfrak{C}$ is completed.

We shall use tensor multiplication when we construct a combinatorial model of the Euclidean space defined as the tensor degree of a one-dimensional complex considered in Sec. 1, Ch. 0. In addition, the extension of tensor multiplication to the case of linear spaces (Sec. 3) will be very important.

It should be pointed out that the tensor multiplication of complexes is immediately extended to the complex which is a conjugate of the complex $\mathfrak{C}$ defined by relations (9)–(11). The complex K

of cochains over a tensor product will automatically also have the structure of a tensor product. The relation

$$d(c_p \otimes c_q) = dc_p \otimes c_q + (-1)^p c_p \otimes dc_q \tag{12}$$

is satisfied for the cochains c_p, c_q and the operator d with the preservation of the property $dd = 0$.

2.5. Concluding remarks. Returning to the remark made at the end of 2.2, concerning the insensitivity of the structure $\mathcal{H}(\mathfrak{C})$ to the refinement of the partition, we shall see now that in fact a much more general statement is valid whose proof goes far beyond the framework of our discussion [21, 30, 31].

THEOREM (Topological invariance of the homology group). *The homology group of a complex associated with a certain geometric object is a topological invariant of the latter.*

This formulation contains such ambiguous terms as a "geometric object" and an "association". In the ultimately general case we understand a "geometric object" simply as a topological space. As to the association of a geometric object with a complex, there are a number of competing procedures here, each of which has its advantages and deficiencies. One of the simplest procedures is the partitioning into cells and the use of a construction similar to that considered in the examples. It is quite correct, in any event, for the situations we have discussed.

In the models we shall be interested in, we shall deal with complexes of a very special structure constructed from cells of different dimensions (in the model considered in Sec. 1, Ch. 0 we must regard a point as a zero-dimensional cell and an interval as a one-dimensional cell). This special structure will make it possible, in particular, to introduce the $*$ operation which will allow us to create a combinatorial imitation of the property of being Euclidean. A subset of a Euclidean space formed by cells or a topological space homeomorphic to this formation will prove to be a geometric object. It will turn out that the homological characteristics of geometric objects appearing in the formulations of the problems of interest to us will be essential in a number of cases, and thus it is useful to bear in mind that they can be found with the use of the scheme given above.

3. Linear Space and Related Structures

3.0. Preliminary remarks. This section contains standard definitions relating to a linear space and linear mappings followed by a number of nonstandard remarks, which are not accentuated, as a rule, in ordinary courses. We understand "related structures" to be a Euclidean property and the structure of spaces defined as outer or tensor products. The concluding subsection contains some definitions and remarks referring to linear representations of groups.

An essential drawback is the absence of the theory of determinants which, of necessity, is assumed to be known.

The basic material can be found in any course of linear algebra; and as to related structures, it is most convenient to study the introductory chapter in [47].

3.1. Initial definitions. A *linear space* E is an Abelian group in which the multiplication by real (complex) numbers is defined which is subject to the natural requirements given below.

If $a, b, \dots$ are numbers, $u, v, \dots$ are elements of a group, then

$$au = ua, \qquad a(bu) = (ab)u, \qquad 1 \cdot u = u,$$

$$a(u + v) = au + av, \qquad (a + b)u = au + bu.$$

It follows from these relations that $0 \cdot u = \mathbf{0}$, $a \cdot \mathbf{0} = \mathbf{0}$, where the bold zero is an element of the group (of a linear space).

The term "vector space" is a synonym of the term "linear space" that is often used, and therefore we shall sometimes call the elements of E *vectors* and the numbers $a, b, \dots$ *scalars*.

In what follows, unless otherwise specified, we assume E to be **real** (i.e., the numbers $a, b, \dots$ to be real).

The elements $u_1, \dots, u_m$ from E are *linearly independent* if the relation

$$a_1 u_1 + \dots + a_m u_m = \mathbf{0}$$

implies that $a_1 = a_2 = \dots = a_m = 0$.

This definition immediately divides the collection of all linear spaces into two classes, namely, the class of *finite-dimensional* spaces (that possess only a finite number of linear independent elements) and the class of *infinite-dimensional* spaces. In this section we shall only deal with **finite-dimensional** spaces E. In this event, the maximal number of linearly independent elements is the *dimension* of E and is denoted by $\dim E$.

If $\dim E = n$, then the collection of n linearly independent elements $\{e_k\}_1^n \subset E$ forms a *basis*: for any $u \in E$ there exists a representation

$$u = \sum_k u^k e_k, \tag{1}$$

where $\{u^k\}_1^n$ is a uniquely defined collection of numbers called the *coordinates* of u in the basis $\{e_k\}$.

REMARK. The collection K_r of cochains with real coefficients introduced in 2.3 is an example of a linear space which is very important for the subsequent discussion. In this event, $\mathcal{H}_k$ turns out to be a linear space and its dimension coincides with the Betti number for $\mathcal{H}^k$.

3.2. Norm, metric, topology. The linear space E (not necessarily finite-dimensional) is *normed* if, for any $u \in E$, the nonnegative real number $\|u\|$ is defined, which is the norm of u, and the following requirements are satisfied:

1. $\|u\| = 0$ implies that $u = 0$.
2. $\|au\| = |a| \, \|u\|$.
3. $\|u + v\| \leq \|u\| + \|v\|$.

By virtue of these requirements, the norm generates, in a natural way, the corresponding **metric**,

$$\rho(u, v) = \|u - v\|,$$

and a normed space is automatically a metric space and, consequently, **topological**.

Two norms $\|\dots\|_1$, $\|\dots\|_2$ are *equivalent* if there exists a constant c such that the inequalities

$$\|u\|_1 \leq c\|u\|_2, \qquad \|u\|_2 \leq c\|u\|_1$$

hold true for any element $u \in E$.

PROPOSITION 1. *In a finite-dimensional linear space any two norms are equivalent.*

This statement immediately follows from the comparison of norms carried out in an arbitrary fixed basis. ■

When a basis is defined, the relation $\|u\| = \sum_k |u_k|$, where $\{u_k\}$ are the corresponding coordinates, gives the standard technique of specifying the norm.

PROPOSITION 2. *A finite-dimensional linear normed space is complete.* ■

It is assumed to be complete in every metric generated by any one of the equivalent norms. In an infinite-dimensional case the completeness is an **additional assumption**. A complete linear normed space is known as a *Banach space*. This term is used, as a rule, only in an infinite-dimensional case.

It follows from Proposition 2 that the topology in E generated by the choice of a fixed norm (and the corresponding metric) does not depend on the indicated choice and can be regarded as a **natural** topology.

Some typical topological characteristics of a finite-dimensional linear space considered in a natural topology must be pointed out. The linear space E itself is not compact. Nevertheless the following statements are valid.

PROPOSITION 3. *Every closed bounded set of a finite-dimensional linear space is compact.* ■

PROPOSITION 4. *A finite-dimensional linear space is locally compact.* ■

In a certain sense a local compactness is precisely a "topological variant" of finite-dimensionality.

In a finite-dimensional case that ensures the existence of a norm, the talk about topological properties looks somewhat far-fetched, but it is not out of place since in contemporary mathematics an important part is played by a huge class of linear topological spaces (infinite-dimensional, of course) which turn out to be nonnormalizable [10].

3.3. Linear maps. As already pointed out, when we study maps of sets that have one or another additional structure, we first of all investigate maps consistent with this structure in a proper sense. Thus, in the consideration of the mapping $f: E_1 \to E_2$ of linear spaces, of the utmost importance are *linear* mappings $L: E_1 \to E_2$ possessing the property that

$$L(au + bv) = aLu + bLv \tag{2}$$

for any $u, v \in E_1$ and any numbers a, b.

REMARK. As in the case of groups (see 2.1), the notation $\mathrm{Hom}\,(E_1, E_2)$ is often used for the set of all linear maps $E_1 \to E_2$.

If $\{e_k\}_1^n$ is a basis in E_1 and $\{\varepsilon_j\}_1^m$ is a basis in E_2, then L is uniquely defined by the specification of the images of the base elements

$$Le_k = \sum_1^m \lambda_k^j \varepsilon_j, \qquad k = 1, \dots, n. \tag{3}$$

The so-called **tensor** symbolics, in which the sign of the sum is omitted but the summation is implied every time when some index is encountered twice, are very convenient when bases are used for specifying and studying linear maps. In this case summation is carried out over the whole set of values of the corresponding index. In these symbolics the coordinate representation (1) and relation (3) have the form

$$u = u^k e_k, \quad Le_k = \lambda_k^j \varepsilon_j, \qquad k = 1, \dots, n, \quad j = 1, \dots, m.$$

In a similar notation, if

$$L: \ E_1 \to E_2, \qquad u \mapsto v, \tag{4}$$

$u = u^k e_k$, $v = v^j \varepsilon_j$, then

$$Lu = u^k Le_k = u^k \lambda_k^j \varepsilon_j = v^j \varepsilon_j, \tag{5}$$

i.e., we have the relation

$$v^j = \lambda_k^j u^k.$$

This is called the *coordinate* notation of the map L in the bases $\{e_k\}$ and $\{\varepsilon_j\}$.

The so-called **metric** formalism gives an alternative language used in the operation with linear maps (again implying the choice of fixed bases). In this formalism the collection of numbers $\{\lambda_k^j\}$ that defines the mapping is associated with a rectangular table, a matrix. At the same time the rule of multiplication of a matrix by a vector (corresponding to the performance of a linear transformation) is defined as well as the rule of matrix multiplication (corresponding to the superposition or multiplication of transformations).

Thus the matrix $\{\lambda_k^j\}$, that defines map (4) in the chosen bases, has m rows and n columns. The vectors u, v are represented by the coordinate columns (i.e., matrices consisting of a single column). Considering the notation of map (4) in the form (5), we see that under this mapping the jth element of the column v results from the multiplication of the elements of the jth row of the matrix by the elements of the coordinate column of the vector u (with a subsequent summation).

This consideration is a special case of the indicated general rule of matrix multiplication. If, besides mapping (4), we are given the mapping

$$T: \ E_2 \to E_3, \qquad v \mapsto w,$$

and a fixed basis in which T is defined by the relations

$$w^l = \tau^l_j v^j, \qquad l = 1, \ldots, q,$$

is also chosen in E_3, then $TL: \ E_1 \to E_3$, $u \to w$,

$$w^l = \tau^l_j v^j = \tau^l_j \lambda^j_k u^k = \mu^l_k u^k,$$

and the element μ^1_k of the matrix $\{TL\}$, that defines the superposition of the maps T and L (their *product*), results from the element-by-element multiplication of the lth row of the matrix $\{T\}$ by the kth column of the matrix $\{L\}$. Obviously the multiplication defined in this way is noncommutative.

Finally, when we study linear maps, we can use an approach which allows us to use only property (2) and particular additional characteristics of the mapping without using the representation of L in some basis. In this event we usually speak of a *linear operator* $L\colon \ E_1 \to E_2$. This viewpoint is most commonly encountered in the theory of infinite-dimensional (functional, for instance, see Sec. 5) linear spaces.

The supposed finite dimensionality of the spaces E_1, E_2 immediately implies the following statement.

PROPOSITION 5. *Every linear map is continuous in a natural topology.* ■

The linear mappings $A: \ E \to E$, i.e., mappings acting in the fixed space E, occupy a significant place. Mappings of this kind define, in a natural way, the $\mathfrak{U}(E)$ *algebra* (they can be added, multiplied by scalars, multiplied together) with an identity element, namely, with an identity map.

The mapping A is *invertible* (or *nondegenerate*) if there exists a mapping A^{-1} such that $AA^{-1} = A^{-1}A = 1$.

The mapping $A\colon\ E \to E$ can obviously also be regarded as the mapping of E into the space E' of the same dimension but different from E. If A is also nondegenerate, then it establishes a one-to-one correspondence between E and E' and is called an *isomorphism* and the corresponding space is said to be *isomorphic.*

PROPOSITION 6. *The isomorphism between finite-dimensional linear spaces can be established if and only if they are of the same dimension.* ■

An inner isomorphism established by the invertible mapping $A: E \to E$ is known as an *automorphism.*

It is useful to note that when the dimensions of the spaces E_1 and E_2 differ, then the mapping $L\colon\ E_1 \to E_2$ are considered to be nondegenerate if

1. it maps E_1 onto the whole E_2 when $\dim E_1 > \dim E_2$.

2. the condition $\dim E_1' = \dim E_1$ is satisfied when $\dim E_1 < \dim E_2$ for the image E_1' of the space E_1 ($E_1' \subset E_2$).

The automorphism A defined, as any linear mapping, by the map of the basis

$$Ae_k = \alpha_k^j e_j,$$

can be regarded as the transition to a new basis $\{e_{k'}\}$. In this case it is often written as

$$e_{k'} = \alpha_{k'}^k e_k, \tag{6}$$

and, since there exists an inverse

$$e_k = \alpha_k^{k'} e_{k'}, \tag{7}$$

the relations

$$\alpha_{p'}^k \alpha_j^{p'} = \delta_j^k = \begin{cases} 1 & \text{for } k = j, \\ 0 & \text{for } k \neq j \end{cases}$$

hold true. Kronecker's symbol (tensor) δ_j^k, which we have repeatedly used, is one of the standard elements in tensor notation.

It is useful to elucidate the dependence of the coordinate representation of the mapping $L:\ E_1 \to E_2$ on the choice of bases. If the relation

$$u = u^{k'} e_{k'},$$

where the bases $\{e_k\}$, $\{e_{k'}\}$ are connected by relations (6), (7), are valid alongside representation (1), then

$$u = u^k e_k = u^k \alpha_k^{k'} e_{k'} = u^{k'} e_{k'},$$

i.e.,

$$u^{k'} = \alpha_k^{k'} u^k. \tag{8}$$

Turning to L and supposing that the change of the basis in E_2 is defined by the relations

$$\varepsilon_{j'} = \beta_{j'}^j \varepsilon_j, \qquad \varepsilon^j = \beta_{j'}^j \varepsilon^{j'},$$

we have

$$Le_{k'} = \alpha_{k'}^k Le_k = \alpha_{k'}^k \lambda_k^j \varepsilon_j = \alpha_{k'}^k \lambda_k^j \beta_j^{j'} \varepsilon_j = \lambda_{k'}^{j'} \varepsilon_{j'},$$

or

$$\lambda_{j'}^{k'} = \alpha_{j'}^k \lambda_k^j \beta_j^{k'}.$$

If we consider a case of an automorphism and use the operator-matrix notation, then, representing (8) as $u' = Au$ and writing a chain of relations

$$Lu = v, \qquad LA^{-1}u' = A^{-1}v', \qquad L'u' = ALA^{-1}u' = v',$$

we get

$$L' = ALA^{-1}.$$

In conclusion, the following significant factor must be pointed out which is specific for the case of a finite-dimensional linear space. The choice of a fixed basis in E defines the *orientation* of E. From the formal point of view an orientation is the function $\sigma(\varepsilon)$ defined on the set of bases and assuming the values ± 1. In order to define it, we must use the concept of the **determinant** $\det A$ of the square matrix A which we shall assume to be known [26, 52]. Having fixed the basis $\{e_k\}$, for any other basis $\{e_{k'}\}$, we can write relations (6), (7) and set

$$\sigma(e') = \text{sign det} A,$$

where $A = \{\alpha_{k'}^k\}$. The definition is correct since $\det A \neq 0$ and sign det A = sign det A^{-1} for an automorphism of A.

3.4. Functionals, the inner product, the Euclidean space. Besides the significant special class of linear mappings $L\colon\ E_1 \to E_2$, which correspond to the case $E_2 = E_1 = E$, a prominent place is occupied by the case $\dim E_2 = 1$, i.e., the case of the mappings $L:\ E_1 \to \mathbb{R}$. We shall thoroughly study mappings of this kind, which are known as *functionals* (or, in more details, *real linear functionals*). We agree to denote them by lowercase letters.

Thus, suppose that

$$l\colon\ E \to \mathbb{R}$$

is a real linear functional. If there is a fixed basis $\{e_k\}$ in E, it is uniquely defined by a system of numbers $l_k = l(e_k)$, $k = 1, \ldots, n$. It immediately follows that the collection of all linear functionals over E forms, in turn, an n-dimensional vector space, which is said to be a **conjugate** of E and is denoted by E^*. This is one of the most important objects in the theory of linear spaces. We could, evidently, use the notation $E^* = \mathrm{Hom}\,(E, \mathbb{R})$ to define it.

The functionals e^k such that

$$e^j(e_k) = \delta^j_k \tag{9}$$

form a natural basis in E^*. Our functional $l \in E^*$, written in this basis, assumes the form $l = l_k e^k$. Let us calculate its value on the element $u = u^k e_k$. (Without any stipulation, we use tensor notation. In this case the difference of the roles played by the superscripts and subscripts in the representation of the elements E, E^* is very important.) We have

$$l(u) = l_k u^j e^k(e_j) = l_k u^j \delta^j_k = l_k u_k.$$

When we regard l as an element of E^*, this result is usually written in the form

$$\langle u, l\rangle = u^k l_k$$

and we speak of the *pairing* of the elements E and E^*. When we simultaneously consider the spaces E, E^* and call the elements of E **vectors**, then the elements of the space E^* are called *covectors*.

This construction has a number of important consequences. First, it automatically associates every mapping $L\colon\ E_1 \to E_2$ with a certain mapping $L^*:\ E_2^* \to E_1^*$ defined by the relation

$$\langle Lu, q\rangle_2 = \langle u, L^*q\rangle_1.$$

The indices show that on the left-hand side we have a pairing of E_2 and E_2^* and on the right-hand side, that of E_1 and E_1^*. This indication is usually omitted.

Second, using some fixed mapping $M: E \to E^*$, we can associate every element $v \in E$ with an element $Mv \in E^*$ having defined the *inner* pairing

$$\rangle u, v \langle = \langle u, Mv \rangle, \tag{10}$$

that defines over E a *bilinear real functional* which associates a real number with every ordered pair of elements u, v.

It turns out that this idea deserves the most thorough study. Fixing the basis $\{e_k\}$ in E and taking the conjugate basis in E^* (see (9)), we can define M by the relations

$$Me_k = g_{kj} e^j.$$

Then

$$Mv = v^k M e_k = g_{kj} v^k e_j$$

and we can write (10) (after the requisite grouping of factors) as

$$\rangle u, v \langle = g_{kj} e^j e_s v^k u^s = g_{kj} v^k u^j.$$

The natural requirement of the **symmetry** of pairing (10), i.e., the equality of the vectors u, v appearing in it, leads the condition

$$g_{ks} = g_{sk}, \tag{11}$$

which (in a situation of this kind) we shall always suppose to be satisfied.

The definition of the space E^* suggests the choice of the natural isomorphism as a privileged mapping $M: E \to E^*$. It is a nondegenerate mapping for which

$$g_{ks} = \delta_{ks} = \begin{cases} 1, & k = s, \\ 0, & k \neq s. \end{cases} \tag{12}$$

Such a choice of $\{g_{ks}\}$ actually has a fundamental importance, and a special part is played by mappings which look like (12) in the sense that they can be reduced to the form (12) by an appropriate choice of the basis (see (8) and the subsequent reasoning). If a mapping of the type we have described is fixed, then the corresponding pairing (10) is called an *inner* (*scalar*) *product* (nondegenerate, positive definite) and the system of numbers $\{g_{ks}\}$ defining it is called a *metric tensor.*

The origination of the last term (we shall consider the general definition of the term "tensor" entering into it in 3.6) is due to the possibility of specifying in this case the norm of the element $u \in E$ by the relation

$$\|u\|^2 = g_{kj}u^k u^j$$

and, consequently, of defining the corresponding metric in E. The satisfaction of the required axioms therewith follows from the validity of the inequality

$$|(u, v)| \leq \|u\| \, \|v\|$$

for the inner product (which is usually denoted by parentheses).

The linear space in which the inner product, specified by (12) in a certain basis, is defined is known as a *Euclidean space.* The corresponding basis possesses the property that $(e_i, e_j) = \delta_{ij}$ for it and is called an *orthonormal basis.*

Supposing that the specification of an inner product defines in E an **additional structure** (turns it into a Euclidean space); we naturally assume that the value of the product for a fixed pair of vectors does not depend on the choice of the basis (the coordinate representation). This is ensured by the corresponding laws of transformation of the systems of numbers $\{u^k\}$, $\{v^k\}$, $\{g_{ij}\}$ when the basis is changed (cf. (8) and the subsequent reasoning). It stands to reason that a similar *invariance* exists for any fixed pairing defined in some coordinate system chosen in E, E^*.

It should be pointed out that pairings in which the mapping M defining them cannot be reduced to the form (12) also deserve attention. As a rule, pairings of this kind are also called a scalar multiplication. However, the corresponding inner product is *indefinite* (and, therefore, degenerate). The following statements describe all possible situations.

PROPOSITION 7. *Every linear mapping* $M : E \to E^*$ (*satisfying condition* (11)) *can be reduced to the form*

$$g_{ks} = \begin{cases} 0, & k \neq s, \\ \sigma(s), & k = s, \end{cases} \tag{13}$$

where $\sigma(s)$ assumes the values 1, -1, 0, *by an appropriate change of the basis.* ■

PROPOSITION 8. *The numbers* $\sum_s |\sigma(s)|$ *(rank) and* $\sum_{s:\,\sigma(s)=1} \sigma(s)$ *(positive inertia index) do not depend on the choice of reduction of the transformation M to the form* (13). ■

The "choice of reduction" means naturally a change of the basis defined by the nondegenerate transformation.

3.5. Subspaces and the exterior multiplication. The subset $E' \subset E$ of the vector space E is called a *subspace* if any linear combination of the elements of E' again belongs to E'.

We have already encountered the term "object" in our previous considerations. For instance, the image $LE_1 \subset E_2$ of the space E_1 under the linear mapping $L: \ E_1 \to E_2$ is a subspace (proper, i.e., noncoincident with the entire E) if $\dim E_1 < \dim E_2$. Another standard technique of forming a subspace is the formation of a *linear span* of a given collection $\{u_\sigma\}$ of the elements of E, i.e., the addition to this collection of all linear combinations of the elements appearing in it. The dimension of the resulting subspace is obviously equal to the number of linearly independent vectors entering into $\{u_\sigma\}$.

The term *hyperplane* (containing a zero element) is a standard geometric synonym denoting a subspace. A straight line and a plane passing through the origin are visual examples in a three-dimensional space.

REMARK. It pays to note that important "special subsets" whose isolation is admitted by the structure of linear spaces include not only subspaces-hyperplanes but also (when a basis is defined) collections of vectors ("points") whose coordinates $(u_1, \ldots, u_n)$ are related as

$$f_p(u_1, \ldots, u_n) = 0, \qquad p = 1, \ldots, m.$$

In this case we usually speak of a *surface* in a linear space. However, as a rule, a systematic study of formations of this kind (surfaces) is conducted in a different axiomatic. In such a situation all points of an n-dimensional **affine space** are considered which are, roughly speaking, simply ordered collections of n elements of some chosen field $\mathcal{F}$ [29]). However, the term a "surface in a Euclidean space" is also often used (see Ch. 2).

When studying various problems connected with subspaces (or with a linear dependence), a special algebraic operation, known as an **exterior multiplication** of vectors, proves to be very useful. From a formal point of view this operation is a specialization (isolation of the skew-symmetric part) of a more general construction, a tensor product [47], defined in 3.6. Nevertheless, it seems to be expedient to carry out an independent consideration of an exterior multiplication. Since up until now the corresponding material has not always been included into the standard courses of linear algebra, our exposition will be less schematic than that in the preceding subsections.

Thus, for the elements of the basis $\{e_i\}_1^n$ of the linear space E we introduce the operation $\wedge$ which is associative, distributive, anticommutative, and permutable with multiplication by scalars (the elements of $\mathbb{R}$). The $\wedge$ operation is called an *exterior multiplication* and the element $e_{k_1} \wedge \ldots \wedge e_{k_r}$ is an *r-vector*.

EXAMPLE. The outer product of two bivectors $ae_1 \wedge e_3$, $be_2 \wedge e_1$ can be written as a 4-vector

$$(ae_1 \wedge e_3) \wedge (be_2 \wedge e_4) = abe_1 \wedge e_3 \wedge e_2 \wedge e_4 = -abe_1 \wedge e_2 \wedge e_3 \wedge e_4.$$

The collection of all r-vectors naturally forms, in turn, a linear space Λ_r whose basis is usually considered to be formed by $\binom{n}{r}$ different r-vectors in whose notation the indices are arranged in the order of increase. It follows from the definition we have introduced that an r-vector is nonzero if and only if all the elements appearing in the notation of its basis are distinct. In this case Λ_1 naturally coincides with E and $\mathbb{R}$ is customarily identified with the one-dimensional space Λ_0 which has the number 1 as its base element. Then $\Lambda = \bigcup_0^n \Lambda_r$ is an example of the so-called *graded algebra* over $\mathbb{R}$ with dimension 2^n.

The following statement shows the relationship between the operation introduced and the linear dependence in $\mathbb{R}$.

PROPOSITION 9. *The relation*

$$u_1 \wedge u_2 \wedge \ldots \wedge u_m = 0$$

is a necessary and sufficient condition for a linear dependence of the vectors $u_1, \ldots, u_m$. ■

When considering subspaces, it is expedient to use the concept of a *decomposable* r-vector, which can be represented as the outer product of the vectors of E (elements of Λ_1); an arbitrary r-vector is a **linear combination** of decomposable vectors.

PROPOSITION 10. *Every r-dimensional subspace $E' \subset E$ is associated with a uniquely defined (with an accuracy to within a scalar factor) r-vector $w \neq 0$ such that $u \in E'$ if and only if $u \wedge w = 0$.* ■

PROPOSITION 11. *Suppose that E', E'' are subspaces of E defined (in the sense of Proposition 10) by the r_1-vector w_1 and the r_2-vector w_2, respectively. The inclusion $E' \subset E''$ is valid if and only if there exists an $(r_2 - r_1)$-vector v such that $w_2 = w_1 \wedge v$.* ■

The reader can find the proof of these statements in [47].

Proposition 10 identifies the subspaces of E with the equivalence classes of r-vectors (differing by a scalar factor). Restricting the equivalence relation by the requirement of **positivity** of the factor, we arrive at the concept of an **oriented** subspace (hyperplane). Every r-dimensional subspace can have exactly two different orientations. This definition of an orientation is consistent with that given at the end of 3.3.

Obviously, the definition of the exterior multiplication can be generalized, without any alterations, to the elements of the space E^*; and then, if $\{e^k\}$ is the basis of E^* which is a conjugate of the basis $\{e_j\}$ in E, then the collection $\{e^{k_1} \wedge \ldots \wedge e^{k_r}\}$ of r-covectors naturally forms a basis of linear functionals over Λ_r. It is sufficient to define the pairing of the base elements by the relation

$$\langle e_{j_1} \wedge \ldots \wedge e_{j_r}, e^{k_1} \wedge \ldots \wedge e^{k_r} \rangle = \langle e_{j_1}, e^{k_1} \rangle \ldots \langle e_{j_r}, e^{k_r} \rangle, \tag{14}$$

extending it by linearity to the arbitrary pair $\langle r$-vector, r-covector$\rangle$. Note that r-covectors are also called *forms of degree* r (or *of order* r).

When E contains an additional structure, an inner product, then a formula similar to (14) defines the inner product of r-vectors. For simple r-vectors $u = u_1 \wedge \ldots \wedge u_r$, $v = v_1 \wedge \ldots \wedge v_r$ we have

$$(u, v) = (u_1, v_1) \ldots (u_r, v_r) = g_{j_1 k_1} \cdots g_{j_r k_r} u_1^{j_1} v_1^{k_1} \ldots u_r^{j_r} v_r^{k_r}. \tag{15}$$

As in the case of vectors, the value of the inner product of r-vectors does not depend on the choice of a coordinate system, and therefore we can establish its properties in an orthonormal basis in which formulas of form (15) are naturally much simpler.

In particular, if an inner product is positive and nondegenerate, then it is easy to verify that we can set

$$(u, u) = \|u\|^2$$

for the r-vector u. Then the necessary axioms will be satisfied. In this case $\|u_1 \wedge \ldots \wedge u_r\|$ gives the volume of the parallelepiped constructed on the vectors $u_1, \ldots, u_r$.

EXAMPLE. Suppose that e_1, e_2 are base vectors of the Euclidean plane and the vectors $u_1 = ae_1 + be_2$, $u_2 = ce_1 + de_2$ define the sides of the corresponding parallelogram. Then

$$u_1 \wedge u_2 = (ad - bc)e_1 \wedge e_2,$$

and the definition of the volume given above gives a standard result.

The relationship between our constructions and the classical theory of determinants and volumes shown in this example remains valid in an arbitrary dimensionality.

The existence of a metric structure (an inner product) allows us to introduce one more important operation

$$* : \Lambda_r \to \Lambda_{n-r}$$

of *metric conjugation* defined for the arbitrary r-vector u by the relation

$$u \wedge *u = \|u\|^2 e_1 \wedge \ldots \wedge e_n,$$

where $e_1, \ldots, e_n$ are base vectors (of an orthonormal basis).

EXAMPLE. Let $\dim E = 3$, $u = u^k e_k$, $v = v^k e_k$ be a pair of vectors. Then

$$*(u \wedge v) = (u^2v^3 - u^3v^2)e_1 + (u^3v^1 - u^1v^3)e_2 + (u^1v^2 - u^2v^1)e_3,$$

i.e., the left-hand side of the relation is the notation of an ordinary vector product with the use of our symbols.

The $*$ operation has a transparent geometric meaning, namely, it associates the element u with an orthonormal element of an additional dimension and the corresponding volume. In the case of a vector product this corresponds to an ordinary statement concerning a "vector orthogonal to the surface element $u \wedge v$ and equal to its area in absolute value".

It is easy to verify that $** = \pm 1$. The sign depends on the relation between n and r. Hence we have the following statement.

PROPOSITION 12. *The mapping $*$ defined an involute (or anti-involute) isomorphism of the spaces Λ_r and Λ_{n-r}.* ■

PROPOSITION 13. *For the r-vectors u, v we have a relation*

$$(u, v) = \langle u \wedge *v, e^1 \wedge \ldots \wedge e^n \rangle,$$

where the symbol $\langle \ldots, \ldots \rangle$ denotes the operation of pairing of the spaces E, E^*. ■

It remains to note that the metric structure in E automatically induces it into E^* as well (a basis which is a conjugate of an orthonormal basis is considered to be orthonormal). Then the norm, the $*$ operator, and the corresponding statements are generalized to the space Λ^r.

3.6. Tensor algebra. Let E_1, E_2 be a pair of vector spaces with bases $\{e_1\}_1^n$, $\{\varepsilon_j\}_1^m$, respectively. Let us define the linear space $E = E_1 \otimes E_2$ of dimension nm, taking the pairs $e_i \otimes \varepsilon_j$ as the basis and postulating the identifications

$$(e_i + e_k) \otimes \varepsilon_j = e_i \otimes \varepsilon_j + e_k \otimes \varepsilon_j, \quad e_i \otimes (\varepsilon_j + \varepsilon_l) = e_i \otimes \varepsilon_j + e_i \otimes \varepsilon_l, \tag{16}$$

$$a(e_i \otimes \varepsilon_j) = ae_i \otimes \varepsilon_j = e_i \otimes a\varepsilon_j, \tag{17}$$

which are evidently similar to (7), (8) from Sec. 2, with the only difference that now a is an arbitrary real number and (17) does not follow from (16). We call the obtained space E a *tensor product* of the spaces E_1, E_2. The remarks made in 2.3 concerning the tensor product of Abelian groups obviously remain valid.

Of especial importance are tensor products connected with the fixed space E, i.e., the tensor **degrees** of E, E^* or products containing E, E^* as factors. The collection of the corresponding products written as

$$\mathcal{T}(E) = \mathbb{R} + E + E^* + E \otimes E + E \otimes E^* + E^* \otimes E + E^* \otimes E^* + \ldots \tag{18}$$

(where $\mathbb{R}$ is a field of scalars) is called a *tensor algebra* of the space E. An element of the product

$$\underbrace{E \otimes \ldots \otimes E}_{r \text{ times}} \otimes \underbrace{E^* \otimes \ldots \otimes E^*}_{s \text{ times}}$$

is called a *tensor of type* (r, s) (r times contravariant and s times covariant). The corresponding collection of numbers $u^{i_1 \ldots i_r}_{j_1 \ldots j_s}$, which defines the tensor in a given basis (in its representation in terms of the base vectors), is called the collection of its *components*. The law of transformation of components of a tensor when the basis is changed follows from its definition and discussion carried out in 3.3.

Algebra (18) contains many important subalgebras, namely, covariant tensors alone or contravariant tensors alone, symmetric tensors (whose components do not depend on the sequential order of indices), or skew-symmetric tensors (whose components change sign when the pair of adjacent indices is commuted). The case of a symmetry with respect to one group of indices and skew-symmetry with respect to another group, etc. is also possible. It stands to reason that these characteristics of tensors do not depend on the choice of a basis.

The exterior algebra that we considered in 3.5 is exactly a subalgebra of skew-symmetric tensors contravariant in the case of r-vectors and covariant in the case of r-covectors.

The number systems $\{g_{ik}\}$, which we used to define an inner product, are a collection of components of a twice covariant symmetric tensor.

The following characteristic of a tensor product is often used.

PROPOSITION 14. *There is a natural one-to-one correspondence between the linear mappings $L: E_1 \to E_2$ and the elements $E_1^* \otimes E_2$.*

PROOF. Let us associate a simple element of form $w^* \otimes v$ with the following mapping of the element $u \in E_1$:

$$(w^* \otimes v) : u \mapsto \langle u, w^* \rangle v$$

(in which $\langle \ldots, \ldots \rangle$ is the pairing of the spaces E_1, E_1^*). Extending this mapping by linearity, we obtain the mapping

$$(\lambda_k^j e^k \otimes \varepsilon_j) : u \mapsto \langle u, \lambda_k^j e^k \rangle \varepsilon_j = u_k \lambda_k^j \varepsilon_j$$

for the arbitrary element

$$\lambda_k^j e^k \otimes \varepsilon_j \tag{19}$$

(written in the basis being used); this is equivalent to the specification of a certain map Lu (in the notation used in 3.3).

Conversely, if L is defined by the matrix $\{\lambda_k^j\}$, it is sufficient to associate it with the element from $E_1^* \otimes E_2$ of the form (19). ∎

When $E_1 = E_2 = E$, Proposition 14 gives an example of the tensor $\{\lambda_k^j\}$ which for one time is covariant and for one time is contravariant.

Here is an equivalent formulation of Proposition 14 which is often used.

PROPOSITION **14′**. *There exists a natural isomorphism*

$$E_1^* \otimes E_2 \sim \mathrm{Hom}\,(E_1, E_2). \blacksquare$$

3.7. Linear representations of groups. Here we shall consider the relationship between the constructions discussed in Sec. 2 of Ch. 0; the corresponding construction from Ch. 4; and certain remarks made in Sec. 2, Ch. 5, with the elements of the representation theory. We recommend that the reader study [42]. If the reader is also interested in the relationship between the "Fourier discrete transformation" and the continual transformation considered from the group angle, then it is useful to use [7].

Suppose that E is a **complex** vector space, $\mathcal{L}(E)$ is the group of its automorphisms (Sec. 2), and G is a finite group (i.e., a group with a finite number of elements). The representation G in E is an arbitrary homomorphism

$$T:\ G \to \mathcal{L}(E), \qquad T(gh) = T(g)T(h). \tag{20}$$

In this case the dimension of the space E is called the *degree* of the representation. If we fix a certain basis $\{e_k\}_1^n$ in E and identify the elements $\mathcal{L}(E)$ with the corresponding matrices, then relation (20) can obviously be written in matrix form.

The representations of T, T' in the spaces E, E' are *equivalent* if there exists an isomorphism $B:\ T \to T'$ which establishes a one-to-one correspondence between them.

EXAMPLE 1. Suppose that X is a finite set, G is a group of all one-to-one mappings of X onto itself, and E is a linear space of complex functions over X. Let $g \in G$, $x \in X$. We set

$$T_g f(x) = f(g^{-1}x), \quad \text{where} \quad T_g \equiv T(g). \tag{21}$$

Since T_g is an automorphism of E and $T_g T_k f(x) = f(h^{-1}g^{-1}x)$, relation (21) defines the linear representation of G in E.

EXAMPLE 2. Every representation of degree 1 is a homomorphism $T:\ G \to \mathbb{C}$. Since, in a finite group, for any element g there exists a degree m such that $g^m = 1$, it follows that for the finite group G the value $T(g)$ must always be a root of unity.

REMARK. The trivial representation $T(g) \equiv 1$ for any $g \in G$ is said to be an *identity representation.*

Suppose now that an inner product is defined in the representation space E. It is *invariant* relative to G if

$$(x, y) = (T_g x, T_g y)$$

for any $x, y \in E$ and any $g \in G$.

PROPOSITION 15. *Let (x, y) be an arbitrary inner product in E. Then the new inner product $(x, y)_I$, defined by the relation*

$$(x, y)_I = \sum_{g \in G} (T_g x, T_g y),$$

is invariant relative to G. ∎

Since we can always define an inner product in the finite-dimensional space E, and, according to Proposition 15, it can be considered to be invariant, the use of an invariant inner product, existing in E, in the study of finite-degree representations, is not connected with the introduction of an additional restriction. We shall use this fact.

The subspace E' of the representation space E is *invariant* relative to G if $x \in E'$ implies that $T_g x \in E'$ for any $g \in G$.

PROPOSITION 16. *Suppose that E' is an invariant subspace of E and E'' is an orthogonal complement of E' relative to the invariant inner product*

$$E = E' \oplus E''. \tag{22}$$

Then E'' is an invariant subspace. ∎

If there is an orthogonal decomposition (22) for E in which E', E'' are invariant subspaces, then the homomorphism T in (20) induces the homomorphisms

$$T' : G \to \mathcal{L}(E'), \qquad T'' : G \to \mathcal{L}(E''),$$

which, in turn, define the representations of G in E', E''. In this case we say that T can be decomposed in the *direct sum* of representations

$$T = T' \oplus T''.$$

A representation is *irreducible* if it does not admit of a nontrivial decomposition in a direct sum (a decomposition for which $E = E \oplus 0$ is trivial).

PROPOSITION 17. *Every representation of a finite degree is the direct sum of irreducible representations.*

It is easy to carry out the proof by induction on the dimension of a representation space. ■

PROPOSITION 18. *Every irreducible representation of the commutative group G has a degree one.* ■

This is the end of our brief excursus to the representation theory.

4. Infinitesimal Operations and Smooth Manifolds

4.0. Preliminary remarks. The considerations in Secs. 2 and 3 were algebraic in nature. In this section we shall consider definitions based on the limiting process, the fact emphasized by the attribute "infinitesimal". The use of a limiting process presupposes that the classes of objects in question possess a topological (metric, as a rule, in our case) structure. We should not infer that the problems formulated in terms of a certain infinitesimal structure are, mainly, also of local nature. On the contrary, we shall later make sure that the most interesting and complicated problems from this branch of mathematics are usually connected with the reconstruction and properties of a global object described by a totality of local characteristics.

First of all, we shall consider mappings of linear spaces which are no longer linear. The construction of a linear (in the neighborhood of a fixed point) approximation of such a nonlinear mapping by a linear one leads to the operation of differentiation.

In 4.2 we shall define differentiable (or smooth) manifold, which is one of the most important objects for the sequel. This is a topological space that is locally (again in the neighborhood of every one of its points) constructed as a linear space. It is supplied with a special additional structure which makes it possible to consider problems of analysis on it. The simplest example is a smooth closed (without boundary) surface in a Euclidean space. (A surface with boundary is associated with a specialization of the concept of a manifold, namely, a manifold with boundary.)

Since, at every one of its points, a smooth manifold is "similar" to a linear space, we can locally define on it the structures considered in Sec. 3. The corresponding constructions lead to vector and tensor bundles and their sections (vector and tensor fields). The latter are generalizations of the familiar concept of a numerical (scalar) function of a point. It turns out, in particular, that the functionals over geometric formations ("surfaces" of some dimension) lying in a manifold are defined by sections of Λ^r-bundles, i.e., by differential forms. The process of constructing functionals leads to the integration operation. This process is briefly described in 4.3.

The following remark seems to be appropriate in connection with the content of 4.4. Whereas in a one-dimensional case it is natural to consider a derivative, which has a clear sense of an "instantaneous velocity", to be a primary object of differential calculus, when we pass to $\mathbb{R}^n$, it becomes more convenient to proceed from the concept of a differential to which the idea of a linear approximation of a mapping leads (see 4.1). Subsection 4.4 contains another alternative approach. It is devoted to the operator d defined on sections, r-forms, as a conjugate of the geometric operation of taking the boundary. The coordinate realization of d again leads to differentiation.

It may be of interest that here the primary infinitesimal operation is the integration arising, as was pointed out, in the process of constructing functionals over geometric objects.

4.1. Differentiation. Suppose that E_1, E_2 are finite-dimensional linear spaces, $\mathfrak{U} \subset E_1$ is an open set, and $f : \mathfrak{U} \to E_2$ is a mapping which is not, in general, linear. The mapping f is *differentiable* at the point $x \in \mathfrak{U}$ if there exists a linear mapping $f'(x) : E_1 \to E_2$ such that

$$f(x + h) = f(x) + f'(x)h + r(h)$$

with

$$\frac{\|r(h)\|}{\|h\|} \to 0 \quad \text{as} \quad \|h\| \to 0.$$

The definition does not, obviously, depend on the norms appearing in it (Proposition 1, Sec. 3). The linear mapping $f'(x)$ is called a *differential* of the mapping f at the point x.

If f is differentiable at every point $x \in \mathfrak{U}$, then the map is defined; it is the function of the point

$$f' : \mathfrak{U} \to \mathrm{Hom}\,(E_1, E_2),$$

the so-called *derivative* of the map f.

EXERCISE. Make all that we have said above to conform to the ordinary definitions of a derivative and a differential in the case, where the dimensions of the spaces E_1, E_2 are equal to unity.

Assuming that Hom (E_1, E_2) is supplied, in turn, with the structure of a linear space (Proposition 14, Sec. 3), we can define $(f')' = f''$, and so on. The map f *belongs to the class* $C^k(\mathfrak{U})$ if it is k times differentiable at every point $x \in \mathfrak{U}$ and the map $f^{(k)}(x)$ is continuous.

SOLUTION OF THE EXERCISE. When $E_1 = E_2 = \mathbb{R}$, it is natural to regard the derivative $f'(x)$ as a **numerical** function, i.e. if e, ε are base vectors in E_1, E_2, then

$$f'(x) : \; e \to f'_x \varepsilon, \qquad x \in \mathfrak{U}, \tag{1}$$

where f'_x is a scalar. The map $f'(x)$ is identified with this scalar. When we use the **graphs** of the maps f, f' in rectangular coordinates, with equal scales along the axes ($|e| = |\varepsilon|$), then f'_x is the slope of the straight line, namely, the graph of mapping (1) (and, at the same time, the slope of the **tangent** to the graph of the map f at the point $f(x)$).

We can also establish a similar connection with the ordinary "coordinate" formulations in the general case. If we fix the bases $\{e_i\}_1^n$, $\{\varepsilon_j\}_1^m$ in E_1, E_2 and represent the map $f \in C^1(\mathfrak{U})$ by the collection of numerical functions $\{f_k(x^1, \dots, x^n)\}_{k=1}^m$ ($\{x^i\}_1^n$ are the coordinates of the point $x \in \mathfrak{U}$ in the basis $\{e_i\}$), then the map-derivative $f'(x)$ will be represented by the **matrix** (see 3.3) whose elements, the numerical functions $\{\frac{\partial f_k}{\partial x^j}\}_{j=1,\dots,n}^{k=1,\dots,m}$, are *partial derivatives* of the functions f_k.

The differentiable mapping f is *nondegenerate* at the point $x \in \mathfrak{U}$ if the mapping f' is nondegenerate (see 3.3).

4.2. Smooth manifolds. Suppose that M is a Hausdorff topological space and E is an n-dimensional ($n < \infty$) vector space. The *chart* on M is a homeomorphism

$$\varphi_\alpha : \; E_\alpha \to M_\alpha,$$

where $E_\alpha \subset E$, $M_\alpha \subset M$ are connected open sets. The collection of charts $\{\varphi_\alpha\}$ is such that $\bigcup_\alpha M_\alpha = M$ is called an *atlas*. The mapping

$$\varphi_\beta^{-1} \varphi_\alpha : \; E \to E, \tag{2}$$

is defined at the points $x \in M_\alpha \cap M_\beta$ and we can say that (2) belongs to a certain class C^k. The atlas $\{\varphi_\alpha\}$ *belongs to the class* C^k if all the maps (2), defined on $\bigcup_\alpha \varphi_\alpha^{-1}(M_\alpha \cap M_\beta)$, belong to this class. Saying that the atlas is *smooth*, we suppose that a priori $k \geq 1$ and is sufficiently large for the constructions carried out to be valid (it is customary to use the term smooth when $k = \infty$).

If a certain basis is fixed in E_α, then the coordinates of the vector $\varphi_\alpha^{-1}x \in E$ are called *local coordinates* of the point $x \in M_\alpha$ (on the corresponding chart).

Suppose that manifolds M, M' with smooth atlases $\{\varphi_\alpha\}$, $\{\psi_\beta\}$, respectively, are defined. The mapping $f: M \to M'$ is *smooth* if all $\psi_\beta^{-1} f \varphi_\alpha$, defined on $\varphi_\alpha^{-1}(M_\alpha \cap f^{-1}M'_\beta)$, are smooth. Thus the definition of the smoothness of the map f is connected with the chosen smooth atlases (M, φ_α), (M', ψ_β).

The smooth atlases (M, φ_α), (M', ψ_β) are *equivalent* if the mappings

$$1: (M, \varphi_\alpha) \to (M, \psi_\beta), \qquad 1: (M, \psi_\alpha) \to (M, \varphi_\alpha)$$

are smooth.

A Hausdorff topological space with a *differentiable structure*, namely, the class of equivalent smooth atlases defined on it, is said to be a *differentiable* (*smooth*) *manifold* (of dimension n if E is n-dimensional).

REMARKS. 1. If M is an n-dimensional smooth manifold, then there exists a smooth injection of M into $\mathbb{R}^{2n+1}$. (The exact formulation can be found in [50], p. 160.)

2. One of the remarkable results of differential topology (which studies differentiable manifolds and their smooth mappings) consists of the assertion that on a 7-dimensional sphere in a Euclidean space there exist 28 different differentiable structures.

The smooth manifold M is *orientable* if we can choose the atlas (M, φ_α) such that for all $x \in M_\alpha \cap M_\beta$ the corresponding determinant $\det(\varphi_\beta^{-1}\varphi_\alpha)'$ is positive. The notation $(\varphi_\beta^{-1}\varphi_\alpha)'$ is used for the matrix whose elements are the values of the partial derivatives of the corresponding mapping at the point x (see the end of 4.1).

Suppose that (M, φ_α) is a smooth manifold with maps defined on open sets of the linear space E. Let us consider the collection (with respect to all α) of direct products $M_\alpha \times E$ and perform the identification

$$(x, u) \sim (x, (\varphi_\beta^{-1}\varphi_\alpha)'u) \tag{3}$$

for any $x \in M_\alpha \cap M_\beta$. In the resulting topological space X (with topology defined by that of M and E) we introduce a differentiable structure by associating every φ_α with the homeomorphism

$$\tilde{\varphi}_\alpha : E_\alpha \times E \to M_\alpha \times E,$$

which is coincident with φ_α on E_α and identical on E. The corresponding collection of maps will be an **atlas** for X. The resulting smooth manifold $T(M)$ is known as a *tangent fibration* over M with the projection

$$p : T(M) \to M, \qquad (x, u) \mapsto x.$$

The linear space (*fiber*) $M_x = p^{-1}x$ is a *tangent space* to M at the point x and the vectors from M_x are *tangent vectors* to M at the point x.

The smooth mapping $f_* : T(M) \to T(M')$ of tangent bundles corresponds to the smooth mapping $f : M \to M'$, i.e., if $x \in M_\alpha$, $u \in M_{\alpha x}$, $f(x) \in M'_\beta$, then

$$f_*(x, u) = (f(x), (\psi_\beta^{-1} f \varphi_\alpha)' u). \tag{4}$$

REMARK. It is useful to note that the definition that we have introduced (as in the case of a derivative) is an accurate formalization of "ordinary" definitions. Thus, if a curve $x = f(t)$, $-1 < t < 1$ is defined on a certain surface M referred to the coordinates $x_1, \ldots, x_n$, then the collection of numbers $(\frac{\partial x_1}{\partial t}, \ldots, \frac{\partial x_n}{\partial t})|_{t=0}$ is a tangent vector to this curve at the point $x(0)$.

Now if the curve $f : I \to M$ is defined as a smooth image of the one-dimensional manifold I and $\psi : (-1, 1) \to I$ is a chart on I, $f(0) \in M_\alpha$, then, in the obvious notations, $\varphi_\alpha^{-1} f \psi = (x_1(t), \ldots, x_n(t))$ and if u is a unit tangent vector to I at the point $\psi(0)$,

$$(\varphi_\alpha^{-1} f \psi)'|_{t=0} u = \left(\frac{dx_1}{dt}, \ldots, \frac{dx_n}{dt}\right)\Big|_{t=0} \in M_{\alpha, x(0)}.$$

The *vector field* $\eta(x)$ on M is a section of its tangent bundle, i.e., the mapping $\eta : M \to T(M)$ such that $p\eta = 1$. The field is **smooth** if the section is smooth.

It should now be pointed out that for the given tangent bundle $T(M)$ at every point $x \in M$, along with the tangent space $M_x \equiv E$ we can consider the conjugate space E^*, and any one of the linear spaces, namely, tensor products containing E, E^*, considered in 3.6 (entering into the tensor algebra $\mathcal{T}(E)$).

If W is a linear space of this kind, then the corresponding object $W(M)$, constructed from the direct products $M_\alpha \times W$ according to the same scheme as $T(M)$, is also a *fiber bundle* (*cotangent* if $W = E^*$ and *tensor* if W is a tensor product of a certain number of E, E^*). The difference is in the laws which are similar to laws (3), (4) for $T(M)$. The mappings $(\varphi_\beta^{-1}\varphi_\alpha)'$, $(\varphi_\beta^{-1} f \varphi_\alpha)'$, that correspond to the law of transformation of the vector coordinates must be replaced by mappings corresponding to the laws of transformation of the coordinates of the covector or tensor (thus, if we speak of the **matrix** $(\varphi_\beta^{-1}\varphi_\alpha)'$ appearing in (3) (cf. 4.1), then, in the case of a covector, it must be replaced by a transposed matrix).

The corresponding **fields** (covector, tensor) can be defined by a complete analogy with vector fields.

REMARKS. 1. Proceeding from the collection of direct products $\{M_\alpha \times W\}$, where W is an arbitrary (not appearing in $\mathcal{T}(E)$) vector space, and acting according to the same scheme that we used for defining a tangent bundle, we get a vector bundle over M [33]. However, this time identifications of the form (3) are defined with the use of a special agreement, namely, a specification of a structure group. In turn, a vector bundle is a special case of foliated space (fiber bundle) [48].

2. One of the most important objects of contemporary mathematics is the so-called Lie group, i.e., a group whose elements are, at the same time, points of a smooth manifold, and the group operation $G \times G \to G$ is a smooth mapping. Although this object is mentioned in Ch. 4, it is not discussed in this book. The reader is referred to [1].

3. It should be pointed out that the long chain of definitions which we have introduced and will complement below is not a "working apparatus" systematically used in the sequel, but is only an element of the "foundations". Nevertheless, I think that it is necessary to give it since, to my mind, the presentations encountered in literature are not always satisfactory.

The definition of a differentiable manifold considered at the beginning of this subsection admits of a modification that leads to an important concept of a manifold with boundary. Here is the description of this modification.

In the vector space E with the basis $\{e_i\}_1^n$ we shall consider a half-space, namely, a set of points whose coordinates satisfy the condition $x_1 \geq 0$. We denote it by $\breve{E}$. Let $\breve{E}_\alpha$ be the intersection (nonempty) $E_\alpha \cap \breve{E}$. We can now obtain the concept that we need by introducing, together with charts, half-charts, i.e., homeomorphisms

$$\breve{\varphi}_\alpha \colon \breve{E}_\alpha \to \breve{M}_\alpha,$$

and generalizing, in a natural way, the definitions and agreements introduced above to this case. In this event, the images of points belonging to the hyperplane $x_1 = 0$ are called *boundary* points of the manifold M and their totality, the *boundary* of M.

Here is one more definition. Suppose that (M, φ_α) is an atlas, N is a subset of M, and E' is a subset of E. The map $\varphi_\alpha : E_\alpha \to M_\alpha$ is *N-regular* if either $M_\alpha \cap N$ is empty or φ_α defines the homeomorphism $E_\alpha \cap E' \to M_\alpha \cap N$. The atlas under consideration is *N-regular* if all of its maps are N-regular; N is a *submanifold* of M if, in the differential structure M, there exists an N-regular atlas. The *dimension* of N is, naturally, the dimension of E'.

Finally, when we localize some constructions on a manifold, it is often convenient to use a *decomposition of unity* [39]. For the given open cover $\{\mathfrak{U}_i\}$ on the manifold M (we usually choose this cover proceeding from an atlas), a decomposition of unity is given by the system of functions $\{\eta_i\}$ of the class C^k that possesses the following properties:

1. $\eta_i \geq 0$, $\sum_i \eta_i(x) = 1$ for any $x \in M$.
2. $\eta_i(x) \equiv 0$ for $x \notin \overline{\mathfrak{U}}_i$ (the closure of $\mathfrak{U}_i$).

If M is noncompact and we cannot do only with a finite cover, then we can use the following requirement: every point $x \in M$ should possess a neighborhood which cuts only a finite number of supports of the functions η_i [39].

4.3. Integration. Here and in 4.4 we have to make the description of the concepts that are of interest to us in an essentially vaguer manner as compared to 4.1 and 4.2. Nevertheless, it is convenient to have such a description for the sequel, at least on an intuitive level.

We can characterize the classical process of integration (the construction of a "definite integral") as a process of constructing a functional that associates the numerical characteristic with an object belonging to a certain class (defining, for instance, a surface area, the mass of a solid, the work done by a force along a given curve, etc.). This functional must satisfy a number of special requirements which uniquely define it, in a certain sense.

You can find the set-theoretic exposition of the theory of measure and integration in [40, 55], for instance. The books [50, 53] are devoted to the "geometric" theory of integration on a manifold (whose brief exposition is given below) which uses the vector-covector pairing for the construction of a real functional over "surfaces" or "volumes". An attempt, successful in certain respects of popularizing the ideas we are interested in, was undertaken in [46].

Thus, suppose that S is an r-dimensional bounded surface lying in the Euclidean space $\mathbb{R}^n$ referred to a fixed system of Cartesian coordinates. At the same time, we regard $\mathbb{R}^n$ as a smooth manifold (defined by the specification of a single map), over which, in particular, the bundles $\Lambda_r(\mathbb{R}^n)$, $\Lambda^r(\mathbb{R}^n)$ are defined.

We say that the surface S is **regular** if it admits of the division into parts $\{S_\nu\}_{\nu\in N}$, each of which can be "replaced" (approximated in some sense) by an r-dimensional parallelepiped defined by the r-vector $s_\nu(x) \in \Lambda_r$, where $x \in S_\nu$. Next we consider a sequence of these partitions $\{S^j_\nu\}_{\nu\in N^j}$ and call the quantity

$$\Delta^j = \max_{\nu\in N^j} |s^j_\nu|,$$

where $|S^j_\nu|$ is the Euclidean r-dimensional volume of the parallelepiped s^j_ν, the *mesh* of the decomposition. We assume that $\Delta^j \leq \Delta^{j+1}$.

Suppose now that ω is a section of the bundle $\Lambda^r(\mathbb{R}^n)$, defined over the points $x \in S$. We compose the "integral sum"

$$I^j(S,\omega) = \sum_{\nu\in N_j} \langle s^j_\nu(x), \omega(x)\rangle, \tag{5}$$

where x is a point we have chosen when defining the r-vector $s^j_\nu(x)$ and the angle brackets denote the vector-covector pairing. Under the corresponding regularity requirements entering in the consideration there exists a limit

$$\lim_{\Delta^j\to 0} I^j(S,\omega) = \int_S \omega \tag{6}$$

independent of the technique of formation of the sums (5), which is an *integral* over S of the differential r-form ω (or, simply, of the r-form ω) as the **section** of the bundle Λ^r is called.

Let us discuss some properties of an integral. If we regard the set of r-forms as a linear space, then the linearity of the functional (6) that we have constructed is obvious. In addition, it possesses the additivity property with respect to the partitions of S. The sum of parallelepipeds-covectors appearing in (5) can serve as an example of an r-dimensional **chain** (cf. the definition of a chain in Sec. 2) belonging to the space of chains considered in the geometric integration theory. These chains, in turn, are elements of a linear space, and the analogy between the pairing in (5) and that discussed in 2.3 is complete. At the same time, we can introduce a certain **topologization** in the linear spaces (infinite-dimensional) of the objects under consideration, and the system of definitions ensures a **continuous dependence** of the values of (6) on S and ω.

In the next subsection we shall use the fact that, under appropriate assumptions, the **converse** statement is also true, i.e., **every** functional $\mathcal{L}(S, \omega)$ provided with the appropriate system of properties can be represented as (6).

Let us now consider the definition of an integral in the case where the role of S is played by a regular (in the sense considered above) bounded r-dimensional submanifold N (maybe with boundary) which lies in an n-dimensional smooth manifold M, $r \leq n$.

Suppose that (M, φ_α) is a fixed N-regular atlas over M and $\omega(x)$ is a specified section of the bundle $\Lambda^r(M)$. We also assume that the n-dimensional linear space E, appearing in the definition of M, is provided with a **Euclidean structure** ($E = \mathbb{R}^n$), and the basis that defines the local coordinates in M_α is **orthonormal** and the support of the section $\omega(x)$ under consideration lies in M_α. We set

$$\int\limits_N \omega = \int\limits_S \omega^*, \qquad S = \varphi_\alpha^{-1}(N \cap M_\alpha), \qquad \omega^*(\xi) = \omega(\varphi_\alpha(\xi)), \qquad (7)$$

where $\{\xi\}$ is the above-mentioned fixed system of Cartesian coordinates in E_α.

The case of the section $\omega(x)$ with an arbitrary support reduces to the case that has been discussed owing to the use of decomposition of unity consistent with the use of an atlas (see the end of 4.2) which makes it possible to represent ω as the sum of sections ω_α whose supports lie in M_α:

$$\omega = \sum_\alpha \omega_\alpha = \sum_\alpha \eta_\alpha \omega. \tag{8}$$

Let us discuss the nature of invariance of integrals (6), (7). In definition (6), the Euclidean structure $\mathbb{R}^n$ allowed us to speak of the Cartesian coordinate system and of the r-dimensional volume of the parallelepiped $s^j_\nu(x)$. When we pass to another coordinate system, the invariance of the value of (6) is ensured by the invariance of the pairing and by the corresponding laws of computing the volume in the new coordinates (cf. Sec. 1, Ch. 2).

We cannot carry out measurements on an arbitrary smooth manifold without an additional structure. In definition (7), we introduced such a technique of measurement by assuming that the local coordinates of the "accidental" atlas (M, φ_α) were Euclidean and orthonormal. However, when the atlas and the corresponding value of (7) are fixed, this value remains unchanged upon the passage to another atlas owing to the respective laws of transformation of the objects appearing in (7). This remark is obviously also true for the general case which uses decomposition (8).

The conventional character of invariance of integration on an arbitrary smooth manifold that we have described does not trouble topologists since it does not affect the invariance of the topological characteristics of a manifold that are defined by means of integrals. Neither does it trouble analysts since they always assume that there is a **Riemannian** structure on M (Ch. 2) which allows us to make the invariance unconditional.

It should be pointed out in conclusion that the standard notation for the form $\omega \in \Lambda^r$ in local coordinates is

$$\omega(x) = \sum_{i_1 < \ldots < i_r} \omega_{i_1 \ldots i_r}(x) \wedge dx^{i_1} \ldots \wedge dx^{i_r}. \tag{9}$$

The outer product of differentials entering into (9) should be regarded as the notation of the corresponding base vector in the space Λ^r over x. Notation (9) implies the **identification** of the point $x \in M_\alpha$ with the point $\varphi_\alpha^{-1} x \in E_\alpha$, which is also denoted by x and has local coordinates $x^1, \ldots, x^n$. We shall also use this identification. The use of differential symbols in (9) for denoting the base element is connected with the role ω plays in the integration process that we have described.

4.4. The Stokes formula and the exterior differentiation. The scrupulous realization of the plan, which we gave schematically in 4.3, is very laborious as we can find out if we consider [50, 53]. For our purpose, an accurate limiting process would be ideal that would lead to a continual case from a sufficiently clear discrete (combinatorial) analog of the corresponding constructions considered in detail in Ch. 3 for a special class of complexes. We shall not make such an attempt in this book.

The so-called **general Stokes theorem**, which is one of the most important formulas of the multidimensional analysis, is a direct consequence of the $\langle$chain, form$\rangle$ duality considered in 4.3. It turns out to be a result of the introduction, in the space of forms, of an operation dual to the "geometric" operation ∂ of "taking the boundary", defined over the "geometric formations" V, S, N, ..., which are volumes, surfaces, submanifolds, etc.

The operation ∂, which is customary and obvious when we deal say with a domain lying in $\mathbb{R}^n$ and possessing a sufficiently regular boundary, is generalized by means of the requisite formalization (generalizing the construction carried out in 2.2) to the topological space of chains generating the mapping

$$\partial\colon\ \mathfrak{C}^{r+1} \to \mathfrak{C}^r, \qquad \partial:\ V \mapsto \partial V = S, \tag{10}$$

where, in the first case, ∂ is a linear mapping of the space of chains; and, in the second, the corresponding operation on the geometric formation, namely, the "limit object" of the space of chains.

If the dimensions V, S are equal to $r+1$, r, respectively, then the existence of the $\langle$chain, form$\rangle$ pairing makes it possible to define the functionals $\langle V, \chi\rangle$, $\langle \partial V, \omega\rangle$, whose properties, in turn, ensure their representability by integrals, i.e.,

$$\langle V, \chi\rangle = \int\limits_V \chi, \qquad \langle \partial V, \omega\rangle = \int\limits_{\partial V} \omega.$$

At the same time, we can be sure that the linear operation ∂ in the space of chains (corresponding to the operation of taking the boundary over limiting geometric objects) must be associated with the conjugate operation

$$\langle \partial V, \omega \rangle = \langle V, d\omega \rangle. \tag{11}$$

Written in integral form, relation (11) yields the Stokes formula

$$\int_{\partial V} \omega = \int_{V} d\omega, \tag{12}$$

where d is a certain linear mapping of the space of forms, the sections of the corresponding bundles:

$$d: \; \Lambda^r(x) \to \Lambda^{r+1}(x).$$

The definition we have considered immediately gives a number of important properties of the operation d. Enumerating these properties, we assume the forms to be sufficiently smooth and the domains to be sufficiently regular.

It immediately follows from (12) that the relation $\partial V = 0$ implies $\int_V d\omega = 0$. The form ω that satisfies the relation $d\omega = 0$ is said to be *closed*; the form χ that can be represented as $\chi = d\omega$ is said to be *exact*. It directly follows from the relation

$$dd = 0, \tag{13}$$

which is a consequence of the principal property of the operation of taking the boundary (the property $\partial\partial = 0$) that the exact form is closed. Is a closed form always exact? This question is of a fundamental importance. The answer is that on a smooth manifold M the number of linearly independent closed r-forms, which are not exact, defines the topological invariant M, namely, the Betti number B_r (cf. 2.2). The following example is well known.

EXAMPLE. In the ring $1 \leq x^2 + y^2 \leq 2$ the 1-form

$$\omega_1 dx + \omega_2 dy = \frac{-ydx + xdy}{x^2 + y^2}$$

is closed but not exact. Here ω_1, ω_2 are obtained as the formal partial derivatives of the expression $\tan^{-1} \frac{y}{x}$, which does not define a smooth single-valued function in the ring. We cannot construct an example of this kind in the disk $x^2 + y^2 \leq 2$.

Since the definition of the mapping d (like the operation ∂) is "intrinsic" in nature, we have a corresponding invariance, in the framework of atlases, relative to the choice of local coordinates. If $\varphi : M \to M'$ is a smooth mapping (in particular, a change of coordinates), ω is a form on M', and $\varphi_* \omega$ is the corresponding form on M, then the diagram

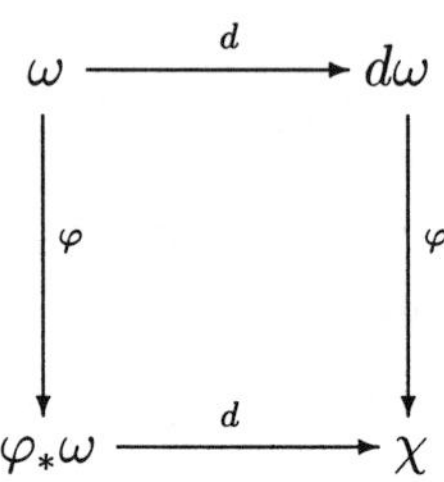

is commutative, i.e., $\chi = d(\varphi_* \omega) = \varphi_*(d\omega)$.

In order to determine the form of the operation d in local coordinates, in the range of ideas of interest to us we should have used a scheme, which we shall illustrate by simple examples without a full realization of it. In the general case, the corresponding considerations require the use of formalism which we shall develop in Ch. 3. For the time being we shall give the general law (16) (see below) without any justification.

Thus, suppose that our original problem of integration is extended to $\mathbb{R}^2$ and the partitioning of the domains of integration reduces to the partitioning of $\mathbb{R}^2$ into squares with the side h, whose vertices lie at the points (ph, qh) (p, q are integers). We assume that the indicated vertices are the base elements of zero-dimensional chains, the oriented intervals (the sides of the squares) are those of one-dimensional chains, and the oriented open squares are the base elements of two-dimensional chains. Suppose that relation (12) is written for the only square V with vertices

$$x_1 = (0,0), \qquad x_2 = (h,0), \qquad x_3 = (0,h), \qquad x_4 = (h,h).$$

The oriented intervals

$$e_1 = (x_1, x_2), \qquad e_2 = (x_1, x_3), \qquad e_1' = (x_3, x_4), \qquad e_2' = (x_2, x_4)$$

are the sides of V (the orientation is in the direction of the coordinate axes). Using the customary notion on the relationship between the positive orientation of V and the orientation of the boundary, we obtain

$$\partial V = e_1 + e_2' - e_1' - e_2. \tag{14}$$

Suppose that in our "local coordinates" the 1-form ω looks like

$$\omega = \omega_1(x)e^1 + \omega_2(x)e^2.$$

If we consider (14) as a 1-chain corresponding to a certain partition of ∂V, then, for a small h, we have

$$\int_{\partial V} \omega \cong h[\omega_1(x_1) - \omega_1(x_3) + \omega_2(x_2) - \omega_2(x_1)]. \tag{15}$$

To obtain (15), we have set

$$\langle e_1, \omega_1(x)e^1 \rangle \cong \omega_1(x_1), \qquad \langle e_1', \omega_1 e^1 \rangle \cong \omega_1(x_3),$$

i.e., have taken the values of the form ω at the initial points of the respective intervals. The factor h is the Euclidean measure of each of the intervals.

Similarly, if the 2-form χ looks like

$$\chi = \chi(x)e^1 \wedge e^2,$$

in the same local coordinates and the square V is associated with the bivector $e_1 \wedge e_2$, then

$$\int_V \chi \cong h^2 \chi(0,0).$$

If now $\chi = d\omega$, then

$$\chi(0,0) \cong \frac{\omega_2(x_2) - \omega_2(x_1)}{h} - \frac{\omega_1(x_3) - \omega_1(x_1)}{h},$$

or

$$d(\omega_1(x)e^1 + \omega_2(x)e^2) = \Big(\frac{\partial \omega_2}{\partial x_1} - \frac{\partial \omega_1}{\partial x_2}\Big) e^1 \wedge e^2,$$

and this corresponds to the classical Green formula. It is customary to write dx^1, dx^2 rather than e^1, e^2, but in this case the symbol of the differential plays only a conventional part.

Using similar arguments, we can easily verify that

$$df = \frac{\partial f}{\partial x_1} e^1 + \frac{\partial f}{\partial x_2} e^2,$$

for the 0-form f (a scalar function), i.e., in this case the operator d coincides with the operator of the gradient.

If $M = \mathbb{R}^3$, then, for the 0-form, the operator d again gives the gradient. If

$$\omega = \sum_1^3 \omega_k e^k, \quad \text{or} \quad \omega = \omega_1 dx + \omega_2 dy + \omega_3 dz,$$

(upon an obvious modification of the notations), then

$$d\omega = (D_x\omega_2 - D_y\omega_1)dx \wedge dy + (D_x\omega_3 - D_z\omega_1)dx \wedge dz + (D_y\omega_3 - D_z\omega_2)dy \wedge dz,$$

i.e., in this case d corresponds to the operator of a curl in vector analysis (with an accuracy to within the numbering of the components). As concerns the operation of divergence, the situation is more difficult. The detailed consideration of the relationship between the formalism we introduce and the classical vector analysis will be given in Sec. 2, Ch. 2.

For the time being, we shall write out the general law of computing the operator d:

$$d\Big(\sum_{i_1<\ldots<i_r} \omega_{i_1\ldots i_r}(x)e^{i_1} \wedge \ldots \wedge e^{i_r}\Big) = \sum_{\substack{i_1<\ldots<i_r\\ k=1,\ldots,n}} D_k\omega_{i_1\ldots i_r}(x)e^k \wedge e^{i_1} \wedge \ldots \wedge e^{i_r}. \tag{16}$$

Assuming that, in accordance with laws mentioned in 3.5, we defined a pointwise external multiplication for forms, we shall note the connection of this multiplication with the operation d:

$$d(\omega \wedge \chi) = d\omega \wedge \chi + (-1)^r \omega \wedge d\chi, \tag{17}$$

where r is the degree of ω. We can verify (17) by a direct calculation, proceeding from (16).

5. Hilbert Space and Differential Operators

5.0. Preliminary remarks. When we consider some special classes of partial differential equations and boundary value problems in

Ch. 2 and discuss a limiting process of transition from a discrete model to a continual case in Ch. 3, we shall use some results from the operator theory in the Hilbert space and the corresponding techniques of reducing boundary value problems to the problems of the operator theory.

In this section I give a brief review of the results that we shall need and, at the same time, elucidate the terms we shall use. The reader can find a more detailed exposition (in accordance with the same scheme) and some more references in [13, 14, 19].

5.1. An abstract Hilbert space. It was noted in Sec. 3 that a complete infinite-dimensional linear normed space is known as a Banach space. Now if the norm is generated by an *inner product*, namely, the numerical function (u, v) of the pair of elements that satisfy the requirements

1. $(u, u) \geq 0$, $(u, u) = 0$ implies that $u = 0$,
2. $(u, v) = \overline{(v, u)}$, where the bar denotes the complex conjugation,
3. $(u + v, w) = (u, w) + (v, w)$, and
4. $(au, v) = a(u, v)$, where a is a scalar,

then the space is known as a *Hilbert space.* In this case, we determine the norm setting $\|u\|^2 = (u, u)$. From the classical (Cauchy–Bunyakovsky–Schwarz) inequality

$$|(u, v)| \leq \|u\| \, \|v\|, \tag{1}$$

which is a consequence of the inner product axioms, it immediately follows that the norm introduced by the indicated method satisfies the requirements enumerated in 3.2.

Unless otherwise specified, the spaces we consider will always be supposed to be real and will be denoted by $\mathfrak{H}$.

We can regard a Hilbert space as an infinite-dimensional Euclidean space, the analogy between finite-dimensional and infinite-dimensional cases playing an important part in the reasoning of an heuristic nature. At the same time, our choice of different ways of defining Euclidean and Hilbert spaces must emphasize a number of essential differences between finite-dimensional and infinite-dimensional cases.

In particular, in an infinite-dimensional case, the technique of introduction of a norm-metric, based on the definition of a metric tensor, is inapplicable in many respects. Although the existence of an inner product allows us to speak of orthonormal systems of elements, an abstractly defined Hilbert space must not, generally speaking, be necessarily *separable*, i.e., possess a **basis** that would admit of a renumbering by a positive integer. (True enough, all concrete functional Hilbert spaces that we shall encounter will be separable.)

Let us consider some more definitions. The subset $\mathfrak{H}'$ of the Hilbert space $\mathfrak{H}$, which, in turn, is a linear space, is a *linear manifold*; and if it is **complete** (in the norm induced by $\mathfrak{H}$), then it is a *subspace*. It stands to reason, that this distinction arises only when $\mathfrak{H}'$ has an infinite number of linearly independent elements. The simplest way of forming linear manifolds is the consideration of the *linear span* $\mathcal{M}$ of a certain fixed subset of elements, i.e., all of its finite linear combinations. Now if $\mathcal{M}$ simultaneously includes the limits of infinite convergent sequences, then the corresponding *closed* linear space automatically turns out to be a subspace. The subset of elements of $\mathfrak{H}$ is *complete* if its linear span $\mathcal{M}$ is *dense*, i.e., its closure coincides with the whole space. The following lemma is very useful.

LEMMA (on an orthogonal decomposition). *Suppose that $\mathcal{M}'$ is a linear manifold and $\mathcal{N}$ is the set of elements $\varphi \in \mathfrak{H}$ such that $(\varphi, y) = 0$ for any $y \in \mathcal{M}'$. Then $\mathcal{N}$ is a subspace of $\mathfrak{H}$, and there exists a unique representation of the form*

$$x = x_{\mathcal{M}} \oplus x_{\mathcal{N}}, \tag{2}$$

where $x_{\mathcal{M}} \in \mathcal{M}$ (the closure of $\mathcal{M}'$), $x_{\mathcal{N}} \in \mathcal{N}$, for any $x \in \mathfrak{h}$. ■

Representation (2) is known as an *orthogonal decomposition* of the element x; the sign $\oplus$ has a corresponding meaning and is also used in the notation $\mathfrak{H} = \mathcal{M} \oplus \mathcal{N}$.

The mappings

$$T: \mathfrak{H} \to \mathfrak{H}, \qquad x \mapsto Tx$$

(the strict requirement is the unambiguous definiteness of the element Tx) are customarily called *operators*. In this case, the *linear* operators play the part of linear mappings considered in 3.3, i.e., $T(\alpha x+\beta y) = \alpha Tx+\beta Ty$. In what follows, unless otherwise specified, all the operators that will be considered are assumed to be linear. In contrast to 3.3, the operator T may turn out to be specified only on a certain linear manifold $\mathfrak{D}(T) \subset \mathfrak{H}$ (the domain of definition) that does not coincide with the whole space, and its norm

$$\|T\| = \sup_{x \in \mathfrak{D}(T)} \frac{\|Tx\|_2}{\|x\|_1}$$

(the norms on the right, in the respective spaces $\mathfrak{H}_k$) is not necessarily finite. The linear operator T is *continuous* ($x_n \to x$ implies that $Tx_n \to Tx$) if and only if it is *bounded* (i.e., its norm is finite).

The consideration of **unbounded** operators in the framework of the theory of normed spaces is a pathological case, in a certain sense, i.e., the definition of the operator is "not consistent" with the main structure. Nevertheless, it turns out that the study of the most important class of operators, generated by differentiation, in the framework of the most "convenient" functional spaces inevitably leads to unbounded operators.

Operators generated by differential operations usually arise as a result of the extension of classical definitions to a wider class of objects. The following definition gives an abstract equivalent of such a procedure.

DEFINITION. The operator $\tilde{T}$ is the *extension* of the operator T ($T \subset \tilde{T}$) if $\mathfrak{D}(T) \subset \mathfrak{D}(\tilde{T})$ and both operators coincide on $\mathfrak{D}(T)$.

The simplest example is an extension by continuity of a bounded operator, defined on a linear manifold, to the corresponding subspace. The closure operation gives another important technique of an extension of an operator.

DEFINITION. The operator $T : \mathfrak{H}_1 \to \mathfrak{H}_2$ is *closed* if the convergence $x_n \to x$, $Tx_n \to f$ implies that $x \in \mathfrak{D}(T)$, $Tx = f$. The minimal closed extension of the operator T (provided that it is exists) is the *closure* of T.

The differential operations $T(D)$, which are initially defined on the manifolds of smooth functions lying in the corresponding Hilbert spaces (see 5.2), admit as a rule of the closure of T. If it turns out that the *inverse operator* T^{-1} ($T^{-1}T = 1$ on $\mathfrak{D}(T)$, the so-called *left* inverse operator) exists, then it is usually bounded. These are precisely the properties of $T(D)$ that make it possible to overcome the difficulties connected with unboundedness.

It should be pointed out in passing that the necessary and sufficient condition for the existence of the inverse T^{-1} of T is, as in a finite-dimensional case, the equality $\mathfrak{N}(T) = 0$, where $\mathfrak{N}(T)$ is the kernel of T.

Again as in a finite-dimensional case, the *conjugate* operator T^* is one of the most important objects connected with T. However, this time, setting $y \in \mathfrak{D}(T^*)$ if there exists an element h such that the equality $(Tx, y) = (x, h)$ is satisfied for any $x \in \mathfrak{D}(T)$, we must note that the definition $T^*y = h$ is correct if and only if $\mathfrak{D}(T)$ is **dense** in the corresponding Hilbert space: only under this condition is the element h uniquely defined.

It is useful to note that every operator, defined as a conjugate one, is automatically closed. It follows immediately from the continuity of the inner product, which, in turn, follows from (1).

Functionals, i.e., linear mappings $\mathcal{L} : \mathfrak{H} \to \mathbb{R}, \mathbb{C}$ into the spaces of real or complex numbers which can always be regarded as Hilbert spaces, are an important special case of operators.

The collection of all bounded functionals over a defined Banach space $\mathfrak{B}$ forms the so-called *conjugate* space $\mathfrak{B}^*$ which plays a significant part in many constructions. To a considerable extent, the special position occupied by the Hilbert space (among Banach spaces) is due to the fact that $\mathfrak{H}^*$ always admits of a natural, in a certain sense, **identification** with $\mathfrak{H}$, i.e., a Hilbert space is "self-conjugate". Here is the corresponding statement.

LEMMA (Riesz). *Suppose that $\mathcal{L}$ is a bounded functional defined over the linear manifold $M' \subset \mathfrak{H}$. Then there exists a unique element $h \in M$ (to the closure of M') such that*

$$\mathcal{L}(x) = (x, h), \qquad \|h\| = \|\mathcal{L}\|.$$

for any $x \in M'$. ■

If M' are all $\mathfrak{H}$, then the correspondence $\mathcal{L} \leftrightarrows h$ gives the indicated identification.

5.2. Functional spaces $\mathbb{H}$ and W. With a rare exception, we shall use, in the main part of the book, two concrete function Hilbert spaces which will now be briefly characterized.

Suppose that V is a bounded domain of the space $\mathbb{R}^n$ with the boundary $\partial V = S$ which satisfies certain minimal requirements of regularity. Let $C(V)$ be a linear manifold of real functions, continuous in the closed domain $\overline{V}$, with the ordinary linear operations of addition (pointwise) and multiplication by a scalar. Let us define the inner product on $C(V)$, setting

$$(u, v) = \int_V uv\,dv, \qquad u = u(x), \quad v = v(x), \quad dv = dx_1 \ldots dx_n.$$

Completing $C(V)$ by the norm generated by this inner product, i.e., adding the limits of all convergent sequences to $C(V)$, we get the Hilbert space $\mathbb{H}(V)$.

The study of the nature of the "ideal elements" added to $C(V)$ as a result of the indicated abstract procedure of completion is an object of the theory of functions. It is known that every element of $\mathbb{H}(V)$ constructed in this way can be identified with the **class** of functions which are Lebesgue integrable and almost everywhere coincident. It follows that, in particular, for the arbitrary element $u \in \mathbb{H}(V)$ there is no sense in speaking of the "value of u at the point $x \in V$". It is essential that $\mathbb{H}(V)$ is a Hilbert space (a "space of functions square integrable with respect to V"), in which the linear manifold $C(V)$ is dense by definition. Therefore, we often speak of the *inclusion* $C(V) \subset \mathbb{H}(V)$ understanding this as follows.

On one hand, every element of $C(V)$ uniquely defines, in an obvious way, the corresponding element of $\mathbb{H}(V)$. On the other hand, we can assign an exact meaning to the statement "the element $u \in \mathbb{H}(V)$ is a continuous function" assuming that it means that the class containing u includes the continuous function $u(x)$ (uniquely defined) with which this class can be identified in all considerations.

REMARK. For the element $u \in C(V) \subset \mathbb{H}(V)$ **it is meaningful** to speak of the value of $u(x)$ at the point $x \in V$.

The inclusion $C(V) \subset \mathbb{H}(V)$ is *continuous* in the sense that for any $u \in C(V)$ we have

$$|u, \mathbb{H}| \le \operatorname{mes} V \cdot |u, C|, \qquad |u, C| = \max u(x), \quad x \in V,$$

where $|u, \mathbb{H}|$ is the norm of u in $\mathbb{H} \equiv \mathbb{H}(V)$ and $\operatorname{mes} V$ is the "volume" of the domain V.

Note the important property of the elements of $\mathbb{H}(V)$ which will be of use to us. We suppose that the element $u \in \mathbb{H}(V)$ is defined over $\mathbb{R}^n$ setting $u = 0$ for $x \notin V$. We set

$$B_\delta(u) = \sup_{|h|\le\delta} \left\{ \int\limits_V |u(x+h) - u(x)|^2 \, dV \right\}^{1/2},$$

where $x = (x_1, \ldots, x_n)$ and $|h|$ is the Euclidean norm of the vector $h = (h_1, \ldots, h_n)$.

LEMMA (continuity in the mean). *For any fixed element $u \in \mathbb{H}(V)$ the number $B_\delta(u)$ tends to zero as $\delta \to 0$.*

Let us pass to the second space that we need, namely, to the space W^1 of functions that have a generalized derivative.

Most often use is made of the space $\dot{W}^1$, which is a variant of the space W^1 and which contains elements subject to an additional homogeneous boundary condition. Let $\dot{C}^1(V)$ be a linear manifold of functions continuously differentiable in V and subject to the condition $u|_s = 0$, $v|_s = 0$, ..., $s = \partial V$. We define on $\dot{C}^1(V)$ an inner product, setting (it is convenient to restrict the discussion to the case $n = 2$)

$$\{u, v\} = \int\limits_V (D_x u D_x v + D_y u D_y v) \, dV. \tag{3}$$

Completing $\dot{C}^1(V)$ by the norm generated by this inner product, we obtain the Hilbert space $\dot{W}^1$.

PROPOSITION 1. *We have a completely continuous inclusion $\dot{W}^1 \subset \mathbb{H}$, and, for the elements $u \in \dot{W}^1$, the inequality*

$$|u, \mathbb{H}| \le c|u, \dot{W}^1|, \tag{4}$$

where $|u, \dot{W}^1|$ is the norm of u in the space $\dot{W}^1(V)$ and the constant c depends only on the domain. The integral $\int_\gamma u^2 \, d\gamma$ is defined along any sufficiently regular curve $\gamma \in V$ for the elements $u \in \dot{W}^1$; in this event,

$$\int\limits_{\partial V} u^2 \, ds = 0. \ \blacksquare \tag{5}$$

The complete continuity of the inclusion means the compactness in $\mathbb{H}$ of the sets bounded in $\dot{W}^1$.

If we remove the homogeneous boundary conditions, we can construct the space W^1 ("without the point") by replacing $\dot{C}^1(V)$ by $C^1(V)$ and adding the inner product (u, v) to definition (3). The compactness of the inclusion and the integrability of the elements of W^1 along the curves are preserved, but relation (5) is obviously violated. Inequality (4) becomes trivial.

When we pass to an n-dimensional case, the integrability along the curves in Proposition 1 is replaced by the integrability along $(n-1)$-dimensional hypersurfaces. No more modifications are required.

There are also some analogs of Proposition 1 when we pass to spaces W^m that are defined according to the same scheme, with the use of inner products of type (3) which contain derivatives of order $m > 1$. The integrability over $(n-1)$-dimensional surfaces is replaced by the integrability over surfaces whose dimension depends both on m and on n. In particular, the following statement is valid.

PROPOSITION 2. *For $2m > n$, the elements of the space W^m are continuous functions.* ■

In this case the value of the element of W^m is defined at every point (on a "zero-dimensional surface"). Thus, for $n = 1$, the elements of W^1 are continuous, the fact that can be immediately verified.

The literature on function spaces "of type W" is quite voluminous; see [10, 43], for instance.

5.3. Differential operators. We shall illustrate the methods of the use of the operator theory for studying boundary value problems by two typical examples [12, 13].

THE POISSON EQUATION. We shall consider the Dirichlet problem for the Poisson equation

$$-\Delta u = f, \qquad u|_{S=\partial V} = 0 \tag{6}$$

in the domain $V \subset \mathbb{R}^n$ satisfying the conditions that ensure the validity of Proposition 1. The *generalized solution* of problem (6) is the element $u \in W^1$ such that the relation

$$\{u, v\} = (f, v) \tag{7}$$

is satisfied for any $v \in W^1$. It is obvious that every classical solution (that possesses the second derivatives inside V and the first derivatives which are continuous up to the boundary) is a generalized solution at the same time. Conversely, every sufficiently smooth generalized solution is a classical one (to verify this fact, it is sufficient to carry out integration by parts on the left-hand side of (7) and transfer, in this way, all differentiations from v to u, and then use the arbitrariness of v). The problem concerning the sufficient conditions imposed on f and V, which ensure the smoothness of the generalized solution, is a difficult one. We do not consider this problem here.

THEOREM 1. *For any $f \in \mathbb{H}$ a generalized solution of problem (6) exists and is unique.*

PROOF. Setting $f = 0$, $v = u$ in (7), we obtain $|u, \dot{W}^1|^2 = 0$, whence follows the uniqueness of the solution.

In order to prove the existence, we must make note of the fact that the inner product on the right-hand side of (7) defines, for the arbitrary $f \in \mathbb{H}$, the linear functional over $\dot{W}^1$:

$$\mathcal{L}_f(v) = (f, v),$$

which is, by virtue of (4), bounded, i.e.,

$$|(f,v)| \leq |f, \mathbb{H}|\,|v, \mathbb{H}| \leq c|f, \mathbb{H}|\,|v, \dot{W}^1|.$$

Consequently, according to the Riesz lemma (5.1), it can be realized as the inner product $\{u, v\}$ for a certain $u \in \dot{W}^1$, and this gives (7). ■

Some additional remarks are due. The Dirichlet problem with the nonhomogeneous boundary condition $u|_S = \varphi$, when S and φ are sufficiently smooth, can be reduced to a homogeneous one in an ordinary way.

If we use the described approach, we encounter more difficulties in the Neumann problem than in that of Dirichlet since in the former we must add the boundary condition $\frac{\partial u}{\partial n}|_S = 0$ to the Poisson equation. Its consideration requires the introduction of the space W_N^1 obtained by isolating in $C^1(V)$ of the submanifold C_N^1 of functions orthogonal to the constants and completing it with respect to the W^1-norm. The generalized solution for $u, v \in W_N^1$ is defined by (7), and the theorem on the existence and uniqueness can be proved under the condition $f \in \mathbb{H}_N$ (the completion of C_N^1 with respect to the $\mathbb{H}$-norm). Some more difficulties arise when we prove the inclusion theorem.

It should be pointed out in conclusion that the scheme we have used is based, in essence, on the employment of the self-conjugation and positiveness of the operator $-\Delta : \mathbb{H} \to \mathbb{H}$ (defined by the introduction of a generalized solution). This scheme admits a generalization to a wide class of elliptic (and "quasielliptic") operators which may be perturbed by the nonsymmetric subordinate part, and is closely connected with variational methods.

THE ACOUSTICS EQUATIONS. This is the term that is often used for the simple hyperbolic system of equations

$$D_t u_1 + D_x u_2 = f_1, \qquad D_t u_2 + D_x u_1 = f_2. \tag{8}$$

We shall consider it in the rectangle $V = [0 \le t \le b_1] \times [0 \le x \le b_2]$ under the boundary conditions

$$u_1\big|_{t=0} = u_2\big|_{t=0} = 0, \qquad u_1\big|_{x=0} = u_1\big|_{x=b_2} = 0. \tag{9}$$

We introduce the notations $u = (u_1, u_2)$, $f = (f_1, f_2)$ and write (8) as the equality

$$Lu = f.$$

The notation $u \in \mathbb{H}^1$, C^1, ..., etc. means that the corresponding space includes every component u. The inner product and the norm for $u, v \in \mathbb{H}$ are defined in the ordinary way, namely,

$$(u, v) = (u_1, v_1) + (u_2, v_2), \qquad (u, u) = |u, \mathbb{H}^2|.$$

DEFINITION. The element $u \in \mathbb{H}$ is called a *generalized solution* (*strong*) of problem (8), (9) if there exists a sequence $\{u_i\}$ of elements, that belong to C^1 and satisfy conditions (9), and is such that

$$u_i \to u, \qquad Lu_i \to f$$

(convergence in $\mathbb{H}$).

This definition specifies the closed operator $L : \mathbb{H} \to \mathbb{H}$. This operator is **nonself-conjugate** and its study in the framework of the operator theory is inevitably connected with the consideration of the transposed (formally conjugate) operator L^1 generated by the problem

$$-D_t v_1 - D_x v_2 = g_1, \qquad -D_t v_2 - D_x v_1 = g_2, \tag{10}$$

$$v_1\big|_{t=b_1} = v_2\big|_{t=b_2} = 0, \qquad v_1\big|_{x=0} = v_1\big|_{x=b_2} = 0. \tag{11}$$

The generalized solution of problem (10), (11) (the corresponding operator is $L^1 : \mathbb{H} \to \mathbb{H}$) is defined in the same way as for problem (8), (9).

Equations (8), (10) are related by the **Green formula**. On the assumption that $u, v \in C^1$, it has the form

$$(Lu, v) = \Big(\int\limits_{t=b_1} - \int\limits_{t=0}\Big)(u_1 v_1 + u_2 v_2)\, dx +$$

$$\Big(\int\limits_{x=b_2} - \int\limits_{x=0}\Big)(u_2 v_1 + u_1 v_2)\, dt + (u, L^t v). \tag{12}$$

If u, v are subject to the boundary conditions (9), (11), then (12) turns into the equality

$$(Lu, v) = (u, L^t v),$$

which gives an ordinary relationship between the given operator and the transposed one.

LEMMA 1. *The inequality (fundamental)*

$$|u, \mathbb{H}| \le c|Lu, \mathbb{H}| \tag{F}$$

holds for the generalized solutions of problem (8), (9).

PROOF. It is, obviously, sufficient to prove inequality (F) for $u \in C^1$ which satisfies conditions (9). We use (12), setting $v = u$ and integrating not over the whole domain V but over the subdomain in which $0 \le t \le \tau \le b_1$. Since $L^t = -L$, we obtain

$$2(Lu, u)_\tau = \int\limits_{t=\tau} (u_1^2 + u_2^2)\, dx.$$

Noting now that

$$|Lu, \mathbb{H}|\, |u, \mathbb{H}| \ge |(Lu, u)_t| = \frac{1}{2} \int\limits_{t=\tau} (u_1^2 + u_2^2)\, dx, \tag{13}$$

and integrating (13) with respect to τ in the limits from 0 to b_1, we get (F). ■

COROLLARIES. 1. *The generalized solution of problem* (8), (9) *is unique.*

2. *The range* $\mathfrak{R}(L)$ *of the operator* $L : \mathbb{H} \to \mathbb{H}$ *is a closed subspace of* $\mathbb{H}$.

The corresponding statements (inequality (F^t) and corollaries 1^t, 2^t) are also valid for the operator $L^t : \mathbb{H} \to \mathbb{H}$ (for problem (10), (11)) since they obviously admit of similar proofs.

DEFINITION. The element $u \in \mathbb{H}$ is a *weak solution* of problem (8), (9) if the relation

$$(u, L^t v) = (f, v)$$

holds for any $v \in \mathfrak{D}(L^t)$.

We can similarly define (with the use of the relation $(v, Lu) = (g, u)$) the weak solution v of problem (10), (11). The definitions we have given yield the following lemma.

LEMMA 2. *The orthogonal complement of the subspace* $\mathfrak{R}(L)$ *in* $\mathbb{H}$ *consists of weak solutions of the equation*

$$L^t v = 0. \blacksquare$$

LEMMA 3, 3^t. *The weak solutions of problems* (8), (9), *and* (10), (11) *are unique.*

Now Corollary 1 of Lemma 1, Lemma 2, and part 3^t of the last lemma immediately imply the following theorem.

THEOREM 2. *For any* $f \in \mathbb{H}$, *the generalized solution of problem* (8), (9) *exists and is unique.*

The similar theorem 2^t is obviously valid for problem (10), (11) as well.

This scheme of reasoning, that essentially used the relation $L^t = -L$, admits of an extension to a considerably more general case of the so-called symmetric first-order positive systems [12, 64], for which $L^t = -L + A$ and the operator A is subject to the corresponding (consistent with L) system of requirements.

In the realization of the scheme we have used, especially difficult is the proof of Lemma 3, 3^t (there are a number of problems in which the attempts to overcome failed). A standard technique is the verification that the weak solution is simultaneously a strong one. Since the converse statement is obvious, the verification is equivalent to the proof of the coincidence of the weak and the strong definition of the operators L, L^t. The apparatus that is used in this case, the so-called **mollifiers** (the term introduced by Friedrichs for averaging operators), will be described in 5.4, where, in addition, the technique of proving Lemma 3, 3^t will be illustrated by a model example.

5.4. Mollifiers. Smoothing makes it possible to associate an element of a function space with an element of the same space which is smoother (more regular) and, at the same time, close to it with respect to the corresponding norm. We shall restrict our discussion to a one-dimensional case.

Suppose that $\xi \in \mathbb{R}^1$ and $\omega(\xi) \in C^\infty$ is an even nonnegative function, $\omega(\xi) = 0$ for $|\xi| \geq 1$, and $\int \omega(\xi)\, d\xi = 1$ (we mean the integration along the whole line). We introduce the notations

$$\omega_\varepsilon = \varepsilon^{-1}\omega, \qquad \omega_\varepsilon(x, x') = \omega_\varepsilon\Big(\frac{x - x'}{\varepsilon}\Big);$$

and obtain

$$\int \omega_\varepsilon(x, x')\, dx = \int \omega_\varepsilon(x, x')\, dx' = 1$$

and $\omega_\varepsilon(x, x') = 0$ for $|x - x'| \geq \varepsilon$.

Suppose that $V = (0, b)$, $u \in \mathbb{H} \equiv \mathbb{H}(V)$, and

$$J_\varepsilon u(x) = \int_V \omega_\varepsilon(x, x')u(x')\, dx', \qquad \varepsilon > 0.$$

We call the integral operator J defined in this way the *standard mollifier*. Here are its properties.

J-1. For any element $u \in \mathbb{H}$ the element $J_\varepsilon u \in C^\infty$, and

$$D_x^m J_\varepsilon u = \int_V [D_x^m \omega_\varepsilon(x, x')]u(x')\, dx'.$$

J-2. For any $\varepsilon > 0$ we have $\|J_\varepsilon\| \leq 1$ (since only the space $\mathbb{H}$ enters into the considerations, we shall use the ordinary notation for the norm).

J-3. As $\varepsilon \to 0$, the norm $\|J_\varepsilon u - u\|$ tends to zero.

To prove this fact, it is convenient to use the property of continuity in the mean of the elements of $\mathbb{H}$ indicated in 5.2.

J-4. If $a(x) \in C$, then

$$\|aJ_\varepsilon u - J_\varepsilon(au)\| \to 0 \quad \text{as} \quad \varepsilon \to 0$$

for any $u \in \mathbb{H}$.

J-5. For any $u, v \in \mathbb{H}$ we have

$$(J_\varepsilon u, v) = (u, J_\varepsilon v).$$

The formula, which is an analog of "integration by parts", plays an important part in applications.

J-6. For any $u, v \in \mathbb{H}$ we have

$$(D_x J_\varepsilon u, v) = -(u, D_x J_\varepsilon v).$$

As for J-5, the proof reduces to a change in the order of integration. The minus sign is a consequence of the oddness of the differentiated kernel $D\omega_\varepsilon(x, x')$.

It is often convenient to use the special mollifiers which possess some additional properties necessary for application. Thus, setting

$$J_\varepsilon^\pm u = \int_V \omega_\varepsilon\Big(\frac{x - x' \pm 2\varepsilon}{\varepsilon}\Big) u(x')\, dx',$$

we have the following.

J-7. For any element $u \in \mathbb{H}(V)$ we have

$$J_\varepsilon^+ u|_{x=b} = 0, \qquad J_\varepsilon^- u|_{x=0} = 0.$$

In this case properties J-1, 2, 3, 4 are preserved for the special smoothings, and instead of J-5 we have the relation

$$(J_\varepsilon^- u, v) = (u, J_\varepsilon^+ v).$$

J-6 can be altered in a similar way.

It is also easy to construct the operator $\dot{J}_\varepsilon$ which possesses the property that $\dot{J}_\varepsilon u|_{x=0} = \dot{J}_\varepsilon u|_{x=b} = 0$ for any $u \in \mathbb{H}$. For this purpose, it is sufficient to take in ω_ε an argument of the form

$$\varepsilon^{-1}[x - (1 - 4b^{-1}\varepsilon)x' - 2\varepsilon].$$

We shall illustrate the use of the operators $J_\varepsilon^\pm$ by applying the scheme described in 5.3 to the elementary Cauchy problem:

$$Lu \equiv D_t u + a(t)u = f, \qquad t \in [0, b], \quad u|_{t=0} = 0 \tag{14}$$

(since we speak of the Cauchy problem, it is natural to pass to the variable t). As was pointed out, we must also consider the problem

$$L^t v \equiv -D_t v + a(t)v = g, \qquad v|_{t=b} = 0. \tag{15}$$

Without repeating the definitions and the stages of reasoning given in 5.3, whose reproduction is obvious, we shall consider the equivalence of the weak and the strong definition of the solution of problems (14), (15) (and this gives the proof of the analog of Lemma 3, 3^t).

PROPOSITION. *The weak solution of problems* (14), (15) *is a strong solution at the same time.*

PROOF. It is obviously sufficient to consider one of the problems. Let u be a weak solution of problem (14). Then, $J_\varepsilon^+ v \in \mathfrak{D}(L^t)$ and $(u, L^t J_\varepsilon^+ v) = (f, J_\varepsilon^+ v)$ for any $v \in \mathbb{H}$, or

$$(u, L^t J_\varepsilon^+ v) = (L J_\varepsilon^- u + \eta_\varepsilon, v) = (J_\varepsilon^- f, v), \tag{16}$$

again for the arbitrary $v \in \mathbb{H}$. Hence

$$L J_\varepsilon^- u = J_\varepsilon^- f - \eta_\varepsilon \to f \quad \text{as} \quad \varepsilon \to 0. \tag{17}$$

The correction η_ε, $\|\eta_\varepsilon\| \to 0$ as $\varepsilon \to 0$, appears as a result of the use of J-4 when we replace $J_\varepsilon^-(au)$ in (16) by aJ^-u. Taking the sequence $\varepsilon_k = 2^{-k}$, $k = 1, 2, \ldots$, we obtain, in accordance with (17), the corresponding sequence $u_k = J_{\varepsilon_k} u$ appearing in the definition of the strong solution. ■

If we had $\alpha(t) D_t u$ in (14) (on the assumption that the equation cannot be divided by $\alpha(t)$), then, in order to obtain (16), we would need one of the most refined properties of smoothing.

J-8 (THE FRIEDRICHS LEMMA). *For any* $u \in \mathbb{H}$ *and the Lipschitz-continuous, piecewise-differentiable (possessing a generalized derivative that belongs to* $\mathbb{H}$*) function* $\alpha(t)$*, we have*

$$\|D_t(\alpha J_\varepsilon u) - D_t J_\varepsilon(\alpha u)\| \to 0 \quad \text{as} \quad \varepsilon \to 0.$$

The proof is usually carried out on the assumption that $\alpha \in C^1$, but is also valid under the assumption that we have made. As distinct from the trivial J-5, we must now change the places of the operations of smoothing and multiplication by the function under the sign of the derivative.

Property J-8 remains valid for special mollifiers.

It remains to be noted that when we pass to a multidimensional case, we take the product-kernel

$$\omega_\varepsilon(x_1, x_1') \ldots \omega_\varepsilon(x_n, x_n').$$

In this case the properties of standard mollifiers and their proofs do not undergo any essential changes.

As concerns special smoothings, they are considerably more difficult. The use of specializations is very diversified and depends on the nature of the boundary conditions which the smoothed element must satisfy. For example, if

$$\Omega_\varepsilon^-(x, x') = \omega_\varepsilon\Big(\frac{x_1 - x_1' - 2\varepsilon}{\varepsilon}\Big)\omega_\varepsilon(x_2, x_2') \ldots \omega_\varepsilon(x_n, x_n'),$$

$$J_\varepsilon^- u(x) = \int_V \Omega_\varepsilon^-(x, x')u(x')\, dx',$$

and the domain V lies in the half-space $x_1 \geq 0$ and has the part S_0 of its boundary lying in the hyperplane $x_1 = 0$ (or close to it), then $J^- u|_{s_0} = 0$.

In order to obtain a smooth function satisfying conditions (9), we must use the combination of the smoothings J^- with respect to t and $\dot{J}_\varepsilon$ with respect to x, and so on. In this event, the technique of the proof of a statement of the type of Lemma 3, 3^t is naturally more complicated.

Chapter 2

Analysis on Riemannian Manifolds

0. Introductory Remarks

From the point of view we have chosen, the multidimensional analysis is an analysis on a Riemannian manifold, and a domain of the Euclidean space is one of the most important cases of the latter. It must be stipulated that the title of the chapter is very pretentious since in the sequel we consider only the simplest elements of the Riemannian structure, not connected with curvature, or covariant differentiation, etc. At the same time, the term "analysis" is understood in a very wide sense since it includes the theory of differential equations and boundary value problems.

However, the elements of the Riemannian structure being introduced prove to be sufficient for defining the classes of partial differential equations. It is natural to study these equations in the above-mentioned context. The structure being described and the equations themselves are objects whose discrete models are considered in Ch. 3. However, I think that the constructions we shall consider and the results that will be obtained deserve attention irrespective of their modeling.

A special place will be occupied by the relationship between the formalism introduced in Sec. 1 and the classical vector analysis and the standard equations of mathematical physics. The latter include the Laplace equation, the wave equation, the Cauchy–Riemann equations and their multidimensional generalizations, the Maxwell equations, and the Navier–Stokes equations.

1. The Riemannian Structure

1.1. The metric tensor. The formal nature of definitions from 1.1, based on 3.4, 3.5 of Sec. 3, Ch. 1, is to some extent mollified by the consideration of the elementary examples such as the polar and spherical coordinates given in 1.2.

Suppose that a bundle $W(M)$, whose fiber is $E^* \otimes E^*$, is defined on the smooth orientable manifold M together with the tangent bundle $T(M)$ with the fiber $T_x = E$ (cf. 3.6, 4.2, Ch. 1). Suppose that, at the same time, we are given a fixed section $W(x)$ of this bundle (continuous, at least), which defines the symmetric, twice covariant tensor field $g_{ik}(x)$. In accordance with the constructions from 3.4, Ch. 1, the tensor $g_{ik}(x)$ induces in every T_x an inner product and a metric. This construction turns M into a *Riemannian manifold with the metric tensor* g_{ik}. As a rule, we additionally assume that this metric is *nondegenerate* (at every point), and most often it is strictly positive definite.

The definition of a metric tensor on M is the introduction of an **additional structure** (Riemannian), which makes it possible to extend essentially the class of objects that admit the invariant (independent of the choice of the atlas) definition. It should be pointed out that such an additional structure, with the strictly positive definite continuous metric tensor, can be defined on every manifold of the class C^2 [48]. This statement is not trivial in the sense that it is not always possible to define, on a manifold of dimension n, a continuous tensor field which has the rank n at every point but a positive index of inertia $n-1$. Such a possibility is equivalent to the existence on M of a continuous nondegenerate section of a tangent bundle, whereas it does not exist on a two-dimensional sphere, for instance.

In what follows, we shall also deal with the Lorentz metric (corresponding to the case described above); however, for the time being, unless otherwise specified, we shall suppose the metric that we shall use to be strictly positive definite.

As was pointed out, the definition of a Riemannian metric is equivalent to the introduction, in every T_x, of a structure of the Euclidean space with all the consequences that were considered in 3.4, 3.5, Ch. 1. In particular, we have an invariant definition of the r-dimensional volume ($r = 1, \dots, n$), which is used in the construction of an integral. This leads to the important possibility of defining the invariant *inner product* of r-forms ($r = 0, 1, \dots, n$) defined by the relation

$$(\omega, \chi)_V = \int\limits_V \omega \wedge * \chi, \tag{1}$$

where $V \subset M$ is a domain. The definition of the operation $*$ was given in 3.5, Ch. 1. Let us consider this definition in the framework of the Riemannian formalism [39]. We denote by

$$e_{1,\dots,n} = \sqrt{\det |g_{ij}|}\, dx^1 \wedge \dots \wedge dx^n$$

the n-form which defines an element of the volume on M, i.e., which is transformed when the atlas is changed according to the law

$$\tilde{e}_{1,\dots,n} = J e_{1,\dots,n},$$

where J is the Jacobian of the transformation of the local coordinates. Recall that the tensor g_{ij} is associated with the contravariant tensor g^{jk} such that $g_{ij} g^{jk} = \delta_i^k$, and the operation of "raising" or "lowering" the indices is defined on the Riemannian manifold. For instance, $\omega^{ij} = g^{ik} g^{js} \omega_{ks}$ (everywhere we use the argument concerning the summation over the repeating indices). Now we have

$$*1 = e_{1,\dots,n}, \qquad (*\,\omega)_{j_1 \dots j_{n-p}} = \pm e_{1,\dots,n} \omega_{i_1 \dots i_p},$$

where $i_1, \dots, i_p$ is a p-tuple of indices that complement $j_1, \dots, j_{n-p}$. We have chosen the sign in the second relation such that for the arbitrary form ω we have

$$*(\omega \wedge * \omega) = \sum_{i_1 < \dots < i_p} \omega_{i_1 \dots i_p} \omega_{i_1 \dots i_p} \geq 0$$

(in the next subsection we shall illustrate these formulas by the examples mentioned above).

The definition of the operator d was given in 4.4, Ch. 1, and need not be commented upon; and the definition of the invariant inner product (1) immediately induces the operator $\delta \colon \Lambda^r \to \Lambda^{r-1}$, the so-called *metric conjugate* of d, i.e., an operator connected with it by the relation

$$(d\omega, \chi)_V = (\omega, \delta\chi)_V, \tag{2}$$

written on the assumption that V is the whole manifold (without boundary) or on the assumption of the finiteness in V of one of the (smooth) forms under consideration. Using the Stokes formula and law (17), Sec. 4, Ch. 1, we obtain

$$(d\omega, \chi)_V = \int_V d\omega \wedge *\chi = \int_V \{d[\omega \wedge *\chi] - (-1)^r \omega \wedge d * \chi\} =$$

$$\int_{\partial V} \omega \wedge * \chi + (-1)^{r+1} \int_V \omega \wedge * [*^{-1} d * \chi] = \int_{\partial V} \omega \wedge *\chi + (\omega, \delta\chi)_V, \tag{3}$$

i.e., $\delta = (-1)^{r+1} *^{-1} d *$. Since $** = \pm 1$, we can, neglecting the sign (we shall do this repeately), set

$$\delta = * d * .$$

The definition immediately yields the identity

$$\delta\delta = 0.$$

The operators d, δ are the initial objects considered in the next section and are fundamental for Chs. 2 and 3. Relation (3), which takes into account the **boundary terms** in contrast to (2), is the **Green formula** that connects the operators d and δ and is very important for the sequel.

The operators d, δ are used for the invariant notation of the Laplace operator

$$-\Delta \equiv (d\delta + \delta d) : \ \Lambda^r \to \Lambda^r$$

(we shall discuss it in greater detail in the next section) and for the notation of multidimensional analogs of the Cauchy–Riemann equations, as well as a number of significant operators of mathematical physics (Sec. 6). The close relationship between d, δ and the classical vector analysis (Sec. 2) allows us to regard a considerable part of the reasoning that follows as a "multidimensional vector analysis".

1.2. Polar and spherical coordinates. Suppose that V is a domain of the Euclidean space $\mathbb{R}^2$ regarded as a noncompact Riemannian manifold with a single map, for which the metric tensor has the form $g_{11} = g_{22} = 1$, $g_{12} = 0$. Let us consider the passage to the polar coordinates setting $0 \in V$,

$$r^2 = x^2 + y^2, \qquad x = r\cos\varphi, \qquad y = r\sin\varphi,$$

and denoting the polar angle by φ. Using the classical relationship between the linear elements, we can write

$$(dx)^2 + (dy)^2 = (dr)^2 + r^2(d\varphi)^2,$$

i.e., in the new coordinates $g_{11} = 1$, $g_{22} = r^2$, $g_{12} = 0$. Correspondingly, we have

$$*1 = e_{12} = r\,dr \wedge d\varphi.$$

If $\omega = \omega_1 dr + \omega_2 d\varphi$, then

$$\omega^1 = g^{1i}\omega_i = \omega_1, \qquad \omega^2 = g^{2i}\omega_i = r^{-2}\omega_2,$$

$$*\omega = -r^{-1}\omega_2 dr + r\omega_1 d\varphi, \qquad **\omega = -\omega.$$

For the Laplace operator applied to the scalar u we get

$$-\Delta u = \delta du = -(*\,d\,*)du = -u_{rr} - r^{-1}u_r - r^{-2}u_{\varphi\varphi}.$$

A similar computation corresponding to the transition to spherical coordinates for $V \subset \mathbb{R}^3$ gives

$$x = r\cos\varphi, \qquad y = r\sin\varphi\cos\vartheta, \qquad z = r\sin\varphi\sin\vartheta,$$

$$(dx)^2 + (dy)^2 + (dz)^2 = (dr)^2 + r^2(d\varphi)^2 + r^2\sin^2\varphi(d\vartheta)^2.$$

We do not carry out all the calculations given for the two-dimensional case but only note that

$$e_{123} = r^2\sin\varphi\,dr \wedge d\varphi \wedge d\vartheta,$$

and, for the Laplace operator applied to the scalar u, have

$$du = u_r dr + u_\varphi d\varphi + u_\vartheta d\vartheta,$$

$$*\,du = (\sin\varphi)^{-1}u_\vartheta\,dr \wedge d\varphi - \sin\varphi\,u_\varphi\,dr \wedge d\vartheta + r^2\sin\varphi u_r\,d\varphi \wedge d\vartheta,$$

$$\Delta u = \delta du = *\,d*du =$$

$$(r^2\sin\varphi)^{-1}\Big\{D_r(r^2\sin\varphi\,u_r) + D_\varphi(\sin\varphi\,u_\varphi) + D_\vartheta\Big(\frac{u_\vartheta}{\sin\varphi}\Big)\Big\}.$$

2. The Orthogonal Decompositions of the Spaces $\mathbb{H}^h$ and the Poisson Equation

2.0. Preliminary remarks. We begin with the operations d, δ and the related operations of a more complicated structure in the case of $\dim M = 3$. This will allow us to establish a direct relationship with the vector analysis and some classical problems of mathematical physics. We shall consider this relationship in detail since the alternative interpretation of standard objects, which uses the formalism of Sec. 1, serves as a starting point for transition to an n-dimensional case. Such a transition, which is of interest in itself, allows us to get a better understanding of the nature of classical constructions.

In 2.3 we shall study the relationship between the analytic characteristics of some of the objects we have introduced and the topological structure of M; and this serves as one of the most vivid examples of the relationship between topology, algebra, and analysis.

2.1. Riemannian formalism and vector analysis. We shall begin the discussion of the vector analysis with remarks that return us to those made at the beginning of 4.3, Ch. 1. In the majority of problems of mathematical physics that use the concept of a vector we deal with objects which are actually forms and which define functionals over formations of a geometric nature.

For example, one of the classical "vector" conditions,

$$\operatorname{div} v = 0,$$

results from the integral relation in which v plays the part of a 1-form, i.e., it originates from the requirement of the equality to zero of the integral $\int_{\partial V} *v$, taken along the boundary of an arbitrary element of the volume V.

Similarly, assuming that the relation $\operatorname{rot}\chi = 0$ implies $\chi = \operatorname{grad}\varphi$ and, correspondingly, the value of the integral

$$\int_\gamma \chi = \varphi(\mathcal{P}_1) - \varphi(\mathcal{P}_2)$$

("the work done by a force in a potential field") depends only on the position of the endpoints $\mathcal{P}_1$, $\mathcal{P}_2$ of the curve γ ($\partial\gamma = \mathcal{P}_2 - \mathcal{P}_1$), we see that $\chi = d\varphi$ again plays the part of a 1-form.

It should be emphasized that the terminology of Ch. 1, which was used for describing the main structures and which will be used in what follows, is conventional now.

Let us proceed with our subject. Suppose that $V \subset \mathbb{R}^3$ is a bounded domain of the Euclidean space with a sufficiently regular boundary regarded as a Riemannian manifold with boundary and with an atlas consisting of a single map. Let $\mathbb{H}^0$, $\mathbb{H}^1$, $\mathbb{H}^2$, $\mathbb{H}^3$ be Hilbert spaces of the forms of the corresponding degree resulting from the completion of the linear manifolds of smooth forms in the metric generated by the inner product (1), Sec. 1:

$$(\omega, \chi)_V = \int_V \omega \wedge *\chi$$

(cf. 5.2, Ch. 1). At the same time, we assume that the differential operations

$$d: \mathbb{H}^k \to \mathbb{H}^{k+1}, \qquad \delta: \mathbb{H}^k \to \mathbb{H}^{k-1}$$

are defined on the smooth forms. It is convenient to suppose that $k = 0, 1, 2, 3$, $\mathbb{H}^{-1} = \mathbb{H}^4 = 0$.

It should now be noted that the elucidation of the correspondence of the operations and relations of vector analysis to the Riemannian formalism is very essentially complicated by the use of the specific features of the dimension $n = 3$ in the first case. Since, due to the indicated isomorphism $\Lambda^r \sim \Lambda^{n-r}$ that generates the isomorphism $\mathbb{H}^r \sim \mathbb{H}^{n-r}$, we can do, in vector analysis, with only two spaces $\mathbb{H}$, $\overline{\mathbb{H}}$ of "scalars" and "vectors":

$$\mathbb{H} \equiv \mathbb{H}^0 \ (\sim \mathbb{H}^3), \qquad \overline{\mathbb{H}} \equiv \mathbb{H}^1 \ (\sim \mathbb{H}^2),$$

and with the triple of operators

$$\text{grad}: \mathbb{H} \to \overline{\mathbb{H}}, \qquad \text{rot}: \overline{\mathbb{H}} \to \overline{\mathbb{H}}, \qquad \text{div}: \overline{\mathbb{H}} \to \mathbb{H}.$$

As was pointed out,

$$\text{grad} \equiv d: \mathbb{H}^0 \to \mathbb{H}^1, \qquad \text{rot} \equiv *d: \mathbb{H}^1 \to \mathbb{H}^1.$$

From the results of 1.1 it immediately follows that an equivalent definition is possible, namely, $\text{rot}\,\omega = \delta * \omega \in \mathbb{H}^1$ ($** = 1$ in this case). The matter with the divergence operator is somewhat more complicated since the two definitions,

$$\text{div} \equiv \delta: \mathbb{H}^1 \to \mathbb{H}^0, \qquad \text{div} \equiv d*: \mathbb{H}^1 \to \mathbb{H}^3,$$

are not distinguished in vector analysis. In the first case the divergence is defined as an operator which is a conjugate of the gradient and which gives the scalar $-D_1\omega_1 - D_2\omega_2 - D_3\omega_3$, and in the second case it is defined as an operator appearing in the Gauss–Ostrogradsky formula (which gives the "flow of the vector ω through the surface ∂V"):

$$\int_{\partial V} *\omega = \int_V d * \omega = \int_V (D_1\omega_1 + D_2\omega_2 + D_3\omega_3)\, dx \wedge dy \wedge dz.$$

In the latter case, in accordance with our rules,

$$* \omega = \omega_3\, dx \wedge dy - \omega_2\, dx \wedge dz + \omega_3\, dy \wedge dz$$

($x^1 = x$, $x^2 = y$, $x^3 = z$). In standard manuals, the minus sign is removed by means of the change in the orientation of the corresponding surface element:

$$-\omega_2\, dx \wedge dz = \omega_2\, dz \wedge dx.$$

The classical relations

$$\operatorname{rot} \operatorname{grad} = 0, \qquad \operatorname{div} \operatorname{rot} = 0$$

are notations of the relation $dd = 0$. In the second case it is given in two versions,

$$\delta * d = * dd, \qquad d * * d = dd.$$

Furthermore, the relations

$$\operatorname{div} \operatorname{grad} = -\Delta = \delta d : \mathbb{H} \to \mathbb{H}, \qquad \operatorname{grad} \operatorname{div} - \operatorname{rot} \operatorname{rot} : \overline{\mathbb{H}} \to \overline{\mathbb{H}}$$

can be represented as

$$-\Delta : \mathbb{H}^0 \to \mathbb{H}^0, \qquad -\Delta = d\delta + \delta d : \mathbb{H}^1 \to \mathbb{H}^1. \tag{1}$$

We should remember that the first of relations (1) can also be written as $-\Delta = d\delta + \delta d$ (in this case $d\delta = 0$), and in general, as was pointed out in 1.1,

$$d\delta + \delta d = -\Delta : \mathbb{H}^k \to \mathbb{H}^k, \qquad k = 0, 1, 2, \ldots. \tag{2}$$

Relation (2), in turn, immediately implies that

$$(d\omega, d\omega) + (\delta\omega, \delta\omega) = \int_V D\omega \wedge * D\omega, \tag{3}$$

for forms of any degree, where the right-hand side is the sum of the squares of the first derivatives with respect to every one of the variables and of every component of ω. In other words, the left-hand side of (3) is an invariant notation of the so-called **Dirichlet integral.**

Recall that the frequently used notation of the vector multiplication $\omega \times \chi$, where ω, χ are covectors, is nothing other than $*(\omega \wedge \chi)$.

2.2. Orthogonal decompositions and the Poisson equation. Having established the relationship between the formalism introduced in Sec. 1 and the vector analysis, and recalling how the latter is related to various notations of the Laplace operator, we shall consider now some special orthogonal decompositions of the Hilbert spaces $\mathbb{H}^r$, which, in turn, are connected with the Poisson equation

$$-\Delta\omega = f. \tag{4}$$

To obviate, for the time being, the additional complications caused by the necessity of taking into account the boundary conditions, we shall suppose that M is a smooth n-dimensional manifold without boundary. Let $\mathbb{H}^r$ be a Hilbert space of forms of degree r resulting from the completion of the linear manifold of the smooth forms with respect to the norm generated by the invariant inner product introduced in 1.1. Let us define the operator

$$d:\ \mathbb{H}^{r-1} \to \mathbb{H}^r$$

as the closure in the $\mathbb{H}$-norm of the corresponding operation specified on smooth forms, i.e., as a strong extension of the operation d (5.3, Ch. 1).

PROPOSITION 1. *The range* $\mathfrak{R}_d \subset \mathbb{H}^r$ *of the operator* d *is a closed subspace in* $\mathbb{H}^r$.

This statement is not trivial. The proof is carried out by methods that will be developed in Secs. 3–5 when we study systems of the first-order equations [14]. We shall not consider it here. ■

It follows from the relation

$$(d\omega, \chi) = (\omega, \delta\chi), \tag{5}$$

which is valid for smooth forms, that

$$\mathbb{H}^r \ominus \mathfrak{R}_d = \mathfrak{N}_\delta^{\mathrm{wk}},$$

i.e., the orthogonal complement of $\mathfrak{R}_d$ coincides with the kernel of the operator $\delta : \mathbb{H}^r \to \mathbb{H}^{r-1}$ understood in the weak sense (5.3, Ch. 1). Since strong and weak definitions of the operators d, δ are always equivalent on a smooth manifold without boundary (this can be easily verified by using the decomposition of unity and smoothings), we obtain

$$\mathbb{H}^r = \mathfrak{R}_d \oplus \mathfrak{N}_\delta. \tag{6}$$

Acting in the same way, but beginning with the operation $\delta : \mathbb{H}^{r+1} \to \mathbb{H}^r$ and using relation (5) again as well as the arguments that we have presented, we get

$$\mathbb{H}^r = \mathfrak{R}_\delta \oplus \mathfrak{N}_d, \tag{7}$$

where $\mathfrak{R}_\delta$, $\mathfrak{N}_d$ are defined by analogy with $\mathfrak{R}_d$, $\mathfrak{N}_\delta$, instead of (6).

Now we note that the relations

$$dd = \delta\delta = 0,$$

which are valid for smooth forms, immediately yield the inclusions

$$\mathfrak{R}_d \subset \mathfrak{N}_d, \qquad \mathfrak{R}_\delta \subset \mathfrak{N}_\delta,$$

and this allows us to write

$$\mathfrak{N}_d = \mathfrak{R}_d \oplus \mathfrak{N}'_d, \qquad \mathfrak{N}_\delta = \mathfrak{R}_\delta \oplus \mathfrak{N}'_\delta.$$

Substituting the first of these relations into (6) and the second into (7), we obtain

$$\mathbb{H}^r = \mathfrak{R}_d \oplus \mathfrak{R}_\delta \oplus \mathfrak{N}_\Delta, \tag{8}$$

where

$$\mathfrak{N}_\Delta = \mathfrak{N}'_d = \mathfrak{N}'_\delta = \mathfrak{N}_d \cap \mathfrak{N}_\delta.$$

The notation $\mathfrak{N}_\Delta$ for $\mathfrak{N}_d \cap \mathfrak{N}_\delta$ implies that this subspace is the kernel of the Laplace operator. The validity of the relation

$$\Delta\omega = 0 \tag{9}$$

for the smooth form $\omega \in \mathfrak{N}_\Delta$ immediately follows from (1). The meaning of relation (9) for the arbitrary element $\omega \in \mathbb{H}^r$, lying in $\mathfrak{N}_\Delta$, must be discussed.

REMARK. For $n = 3$, $r = 1$ relation (8), considered in the finite domain $V \subset \mathbb{R}^3$, is known as the **Weyl–Sobolev decomposition**

[6, 44]. In this case the elements $\mathcal{D}(d)$, $\mathcal{D}(\delta)$ are subject to the homogeneous boundary conditions (see 5.2 below), and relation (8) can be formulated as follows: "the vector-element of the space $\mathbb{H}^1$ can be represented as the sum of a potential vector, a solenoidal vector, and a harmonic vector".

Returning to the consideration of relation (9), we change the plan used in 5.3, Ch. 1 (referring to the Poisson scalar equation), with due account of the fact that we act not in the domain V of the Euclidean space, but on the smooth manifold M and consider differential forms instead of scalar functions.

We introduce a new inner product on $\mathbb{H}^r \cap C^1$ setting

$$\{\omega, \chi\} = (d\omega, d\chi) + (\delta\omega, \delta\chi) + (\omega_\Delta, \chi_\Delta), \tag{10}$$

where ω_Δ, χ_Δ are the projections of ω, χ onto $\mathfrak{N}_\Delta$. We denote by W the completion of the indicated linear manifold with respect to the norm $\{\omega, \omega\} = |\omega, W|^2$, generated by (10) (we should have written W^r, but we omit this superscript).

Suppose now that $\omega \in \mathbb{H}^r \cap C^2$ and satisfies the relation

$$-\Delta\omega = (d\delta + \delta d)\omega = f \in \mathbb{H}^r. \tag{11}$$

Then the relation

$$(d\omega, d\chi) + (\delta\omega, \delta\chi) = (f, \chi) \tag{12}$$

is satisfied for any $\chi \in W \cap \mathbb{H}^r$ (we have no theorem on the inclusion $W \subset \mathbb{H}^r$) (we have taken into account that $f \perp \mathfrak{R}_\Delta$). We say that the element $\omega \in W \cap \mathbb{H}^r$ is the *generalized solution* of (11) if (12) is satisfied for any $\chi \in W \cap \mathbb{H}^r$. We introduce the notation $\mathfrak{R}_\Delta = \mathfrak{R}_d \oplus \mathfrak{R}_\delta$.

PROPOSITION 2. *The requirement $f \in \mathfrak{R}_\Delta$ is the necessary condition for the existence of a generalized solution of equation* (11).

Indeed, if $f_\Delta \neq 0$, then, setting $\chi = f_\Delta$, we arrive at a contradiction. ■

Before formulating the uniqueness theorem, we must make note of the following statement.

PROPOSITION 3. *If $\omega \in W \cap \mathbb{H}^r$, $|\omega, W| = 0$, then $|\omega, \mathbb{H}^r| = 0$.*

PROOF. Let us use the representation $\omega = d\psi \oplus \delta\varphi \oplus \omega_\Delta$. It follows from the relation $\{\omega, \omega\} = 0$, in particular, that $\delta d\psi = 0$, i.e., $(d\psi, d\psi) = 0$ (we can use the smooth approximation). In the same way we can establish the relation $|\delta\varphi, \mathbb{H}^r| = 0$. The relation $\omega_\Delta = 0$ is obvious. ∎

PROPOSITION 4. *Under the additional condition* $\omega \in \mathfrak{R}_\Delta$, *the generalized solution of* (11) *is unique.*

Indeed, in this case the relation $-\Delta\omega = 0$ *implies* $|\omega, W| = 0$. ∎

The fact that $\mathfrak{R}_\Delta$ appears in the solvability condition and in the condition of uniqueness at the same time is a natural consequence of the formal **self-conjugacy** of the operator Δ.

It remains to use decomposition (8) for proving the existence of the generalized solution for the arbitrary $f \in \mathfrak{R}_\Delta$. Let us represent the right-hand side of (11) as

$$f = f_d \oplus f_\delta, \qquad f_d = df_1, \quad f_\delta = \delta f_2.$$

In (11) we must obviously have

$$d\delta\omega = f_d, \qquad \delta d\omega = f_\delta.$$

We can solve the equation $d\delta\omega_1 = df_1$, or $\delta\omega_1 = f_1$, under the additional assumption that $d\omega_1 = 0$ so that $\delta d\omega_1 = 0$. We do the same to the equation $\delta d\omega_2 = \delta f_2$ setting $d\omega_2 = f_2$, $\delta\omega_2 = 0$. We have

$$(d\delta + \delta d)(\omega_1 + \omega_2) = f.$$

We have no reason to state that $\omega_1 + \omega_2 \in \mathbb{H}^r \cap C^2$, but by construction $\omega_1 + \omega_2 \in W \cap \mathbb{H}^r$ and is a generalized solution of Eq. (11) belonging to $\mathfrak{R}_\Delta$. We have the following theorem.

THEOREM. *For any element* $f \in \mathfrak{R}_\Delta$ *there exists a unique generalized solution of Eq.* (11) *belonging to* $\mathfrak{R}_\Delta$.

It is clear now that every form $\omega \in \mathfrak{N}_\Delta$ is a generalized solution of Eq. (9) and it is classical on the smooth M.

It follows from our considerations that for the closed operator $\Delta : \mathcal{D}(\Delta) \cap \mathfrak{R}_\Delta \to \mathfrak{R}_\Delta$ there exists an inverse Δ^{-1} defined throughout the subspace $\mathfrak{R}_\Delta$. Consequently, in accordance with the Banach theorem [10], the following statement is valid.

PROPOSITION 5. *The operator* $\Delta^{-1} : \mathfrak{R}_\Delta \to \mathfrak{R}_\Delta$ *is bounded.* ∎

It should be pointed out in conclusion that the inclusion theorem

$$W \subset \mathbb{H}^r, \qquad |\omega, \mathbb{H}^r| \leq C|\omega, W|,$$

is in fact also valid on M, but we managed to do without it. To obtain it, additional constructions are needed. (Cf. (7), Sec. 3 of this chapter.)

2.3. Theorems of de Rham, Hodge, and Kodaira. Let us return to decomposition (8) and establish the relationship between the constructions we have presented and the algebraic and geometric constructions considered in Sec. 2, Ch. 1. It is useful to bear in mind the point of view concerning the integration and the operation d presented in 4.3, 4.4, Ch. 1.

We shall regard $\mathbb{H}^r$ as a space of **cochains** of some complex $\mathcal{K}(M)$ and $d: \mathbb{H}^r \to \mathbb{H}^{r+1}$ as a **coboundary operator**. Note that when we have a Hilbert structure, the subspace $\mathfrak{N}'_d = \mathfrak{N}_d \ominus \mathfrak{R}_d$ considered in 2.2 is equivalent to the quotient space $\mathfrak{N}'_d = \mathfrak{N}_d / \mathfrak{R}_d$, i.e., $\mathfrak{N}'_d$ is an analog of the r-dimensional **cohomology group** $\mathcal{H}_r(\mathcal{K})$ of the complex $\mathcal{K}(M)$.

It stands to reason that at the level of our constructions this remark is no more than a certain analogy. Nevertheless, the following theorem is valid.

THEOREM (de Rham). *Suppose that on the smooth manifold M we are given a polyhedral subdivision that allows us to associate it with the complex $K(M)$ of cochains (real). Then the cohomology groups $\mathcal{H}_r(K(M))$ and $\mathcal{H}_r(\mathcal{K}(M))$ are isomorphic.* ■

The complexes constructed in 2.2, Ch. 1, for a circle, a sphere, and a torus can serve as examples of complexes defined by a polyhedral subdivision of a smooth manifold.

It follows from this theorem that the dimension of the subspace $\mathfrak{N}'$ is equal to the corresponding **Betti number**. Note that this number is always finite for a smooth compact manifold.

Not only the groups $\mathcal{H}_r(K)$, $\mathcal{H}_r(\mathcal{K})$ are isomorphic for the given M, but also homology **rings** mentioned in Sec. 2, Ch. 1. In this case, the multiplication operation in K is given by the Whitney multiplication (Ch. 3) and in $\mathcal{K}$, by the exterior multiplication of differential forms.

Although the de Rham theorem does not directly enter into the main exposition, it plays a fundamental part in the elucidation of the principal aim of this monograph. The theorem states that some important characteristics of such an object as a differentiable manifold can be obtained as a result of the study of the continual structure defined on it, on one hand, and as a result of the analysis of the corresponding "finite model" of the algebraic character specified by the polyhedral subdivision, on the other.

It should be noted that this is the simplest version of the de Rham theorem which admits of far-fetched generalizations ([38], for instance).

The possibility of using the identification $\mathfrak{N}' = \mathfrak{N}_\Delta$ in (8) constitutes the gist of **Hodge's theorem**, which states that we can take, as the **basis** of the space $\mathcal{H}_r(\mathcal{K})$, forms which are **harmonic** [39, 65].

Decomposition (8) itself for the arbitrary smooth manifold M is known as the **Kodaira theorem.**

In the sequel we shall consider the modification of the results of Sec. 2 to the case of a manifold with boundary. It is precisely the case that is important for the study of boundary value problems for partial differential equations.

3. First-Order Invariant Systems

3.0. Preliminary remarks. This title means that we shall speak of systems of differential equations with first-order partial derivatives written with the use of the operators d and δ. These systems are closely connected with the Poisson equation and with the orthogonal decompositions considered in Sec. 2. We understand invariance in the sense that the form of the notation is preserved after any change of coordinates (with due regard of the transformation of the metric tensor induced by the change).

Since, as before, the considerations are on a manifold without boundary, the attention is focused, to a considerable degree, on the formally algebraic structure of the equations. The investigation of solvability is based on decomposition (8), Sec. 2.

3.1. Classical invariant systems. The attribute "classical" means that in this subsection we assume the dimension of M to be equal to two or three, and, as a result, deal with objects well known in the Euclidean case.

We have already encountered the simplest system of the form of interest to us, namely, the homogeneous equations

$$d\omega = 0, \qquad \delta\omega = 0, \qquad \omega \in \mathbb{H}^1,$$

where d, δ are the closures of the corresponding differential operations (see 2.2). The solutions are the elements of $\mathfrak{R}_\Delta$ (harmonic forms).

Let us consider the corresponding inhomogeneous equations

$$d\omega = f \in \mathbb{H}^2, \qquad \delta\omega = h \in \mathbb{H}^0, \tag{1}$$

assuming that $\dim M = 3$. System (1) is obviously overdetermined (we have three unknown functions-components ω but four equations). The requirement

$$df = 0$$

is an additional necessary condition of solvability of (1). Decomposition (8), Sec. 2, applied to $\mathbb{H}^2$ shows how we can obviate the overdetermination by making system (1) solvable for any $f \in \mathfrak{R}_\Delta \subset \mathbb{H}^2$. It is sufficient to introduce the additional unknown function $\varphi \in \mathbb{H}^2$ and write

$$d\omega + \delta\varphi = f, \qquad \delta\omega = h \tag{2}$$

instead of (1). For the second equation in both (1) and (2) the condition $h \in \mathbb{R}_\Delta \subset \mathbb{H}^0$ is also necessary for solvability (in this case $\mathfrak{R}_\Delta = \mathfrak{N}_d$). Finally, decomposition (8), Sec. 2, and the results of 2.2 give the following statement.

PROPOSITION 1. *For any $f, h \in \mathfrak{R}_\Delta$ there exists a unique generalized solution of system* (2) *which satisfies the conditions $\omega_\Delta = \varphi_\Delta = 0$.* ■

The fact that the solution is generalized is connected with the use of the extensions of the operations d and δ defined in 2.2.

In the notations of vector analysis (2.1), system (2) is equivalent to the system

$$\begin{aligned} \operatorname{rot}\omega + \operatorname{grad}\varphi &= f, \\ \operatorname{div}\omega &= g, \end{aligned} \tag{3}$$

studied by many authors [3, 66].

Let us now study certain properties of system (2) which are very essential for generalizations. Note that the collection (ω, φ) of **odd**-degree forms enters as unknowns into (2) and the collection (f, h) of **even**-degree forms as the right-hand sides. Introducing the notation $(\omega, \varphi) = \omega_{\mathrm{I}}$, $(f, h) = f_{\mathrm{II}}$, we write (2) in the symbolic form

$$L\omega_{\mathrm{I}} = f_{\mathrm{II}} \tag{L}$$

and consider the transposed (formally conjugate) system containing the operator L^t defined by the relation

$$(L\omega_{\rm I}, \chi_{\rm II}) = (\omega_{\rm I}, L^t\chi_{\rm II}),$$

where $\chi_{\rm II}$ is a collection of sufficiently smooth even-degree forms. The system

$$L^t\chi_{\rm II} = g_{\rm I} \tag{L^t}$$

can be written out as

$$d\chi_{(0)} + \delta\chi_{(2)} = g_{(1)}, \qquad d\chi_{(2)} = g_{(3)}, \tag{4}$$

where the subscript in parentheses is the degree of the corresponding form.

In turn, Eqs. (4) can be regarded as an analog of (3). A statement, analogous to Proposition 1, that refers to (2) is obviously valid for them. We shall formulate the final result for systems (L) and (L^t) using the notations whose meaning is obvious.

THEOREM 1. *For any $f_{\rm II}, g_{\rm I} \in \mathfrak{R}_\Delta$ there exists a unique generalized solution of systems (L), (L^t) that satisfies the condition*

$$\omega_{\rm I}|_\Delta = \chi_{\rm II}|_\Delta = 0. \blacksquare$$

Let us now consider the case $\dim M = 2$ emphasizing specially the problem of "two-dimensional vector analysis". Proceeding from our previous considerations, it is easy to indicate two-dimensional analogs of systems (L), (L^t). They can be written out as

$$\begin{aligned} d\omega_{(1)} &= f_{(2)}, \\ d\omega_{(0)} + \delta\omega_{(2)} &= f_{(1)}, \\ \delta\omega_{(1)} &= f_{(0)}, \end{aligned} \tag{5}$$

(instead of ω, f, χ, g we now use the notations ω, f for the collection of even as well as odd forms). In the Euclidean case, Eqs. (5) are homogeneous systems of equations of Cauchy–Riemann (differing by sign). If we assume that the components $\omega_{(1)}$ are the pair of functions ω_1, ω_2, then the first system in (5) will have the form

$$D_x\omega_2 - D_y\omega_1 = f_{(2)}, \qquad -D_x\omega_1 - D_y\omega_2 = f_{(0)}, \tag{L}$$

and the second, the form

$$D_x\omega_{(0)} + D_y\omega_{(2)} = f_1, \qquad D_y\omega_{(0)} - D_x\omega_{(2)} = f_2. \tag{L^t}$$

Thus, in the range of ideas that we use, systems (2), (4) are three-dimensional analogs of the Cauchy–Riemann equations. This was noted in [3, 66]. We shall consider some of the results, connected with this analogy, in an n-dimensional case.

The corresponding analog of Theorem 1 is obviously valid for systems (5). Note that our generalized solutions of the systems (L), (L^t) are actually "almost classical", i.e., for any right-hand sides from $\mathbb{H}^r$ they belong to the corresponding spaces W defined in the same way as in 2.2. We shall consider it in greater detail in the n-dimensional case.

Let us discuss now two-dimensional analogs of the operators of vector analysis. Whereas the operators

$$d\omega_{(0)} = (D_x\omega_{(0)}, D_y\omega_{(0)}),$$

$$-\delta\omega_{(1)} = D_x\omega_1 + D_y\omega_2$$

taken as the gradient and the divergence do not cause any doubt, the answer to the question concerning the analog of a rotor is not so unambiguous. We can take as the "rotor" the operator $d: \mathbb{H}^1 \to \mathbb{H}^2$,

$$d\omega_{(1)} = D_x\omega_2 - D_y\omega_1,$$

and the operator $\delta: \mathbb{H}^2 \to \mathbb{H}^1$,

$$\delta\omega_{(2)} = (D_y\omega_{(2)}, -D_x\omega_{(2)}).$$

Thus, in a two-dimensional case we inevitably have all **four** variants of the operators d, δ (whose number for $\dim M = n$ is $2n$). As was already pointed out, the possibility to manage with three operators in the case $\dim M = 3$ is connected with the use of the duality

$$\mathbb{H}^0 \sim \mathbb{H}^3, \qquad \mathbb{H}^1 \sim \mathbb{H}^2,$$

considered in 2.1.

3.2. The Cauchy–Riemann multidimensional equations and the index. In the same way as we passed from systems (2), (4) to the corresponding two-dimensional systems in 3.1, we shall now introduce their n-dimensional analogs which we shall call the *Cauchy–Riemann multidimensional systems.* For an even n, in the notations used in (5), we have

$$\begin{aligned} \delta\omega_{(1)} &= f_{(0)}, \\ d\omega_{(1)} + \delta\omega_{(3)} &= f_{(2)}, \\ &\dots \\ d\omega_{(n-3)} + \delta\omega_{(n-1)} &= f_{(n-2)}, \\ d\omega_{(n-1)} &= f_{(n)}, \end{aligned} \tag{L}$$

$$\begin{aligned} d\omega_{(0)} + \delta\omega_{(2)} &= f_{(1)}, \\ d\omega_{(2)} + \delta\omega_{(4)} &= f_{(3)}, \\ &\dots \\ d\omega_{(n-2)} + \delta\omega_{(n)} &= f_{(n-1)}, \end{aligned} \tag{L^t}$$

where $\omega_{(r)}$ is the collection of components of a form of the corresponding degree. For an odd n, the form of the last row in each of the systems is different:

$$d\omega_{(n-2)} + \delta\omega_{(n)} = f_{n-1} \quad \text{for} \quad (L),$$

$$d\omega_{(n-1)} = f_{(n)} \quad \text{for} \quad (L^t).$$

Each of the systems contains 2^{n-1} unknown functions (the components of the forms) and the same number of equations.

Using the fact that $(d+\delta)(d+\delta) = d\delta+\delta d = -\Delta$, we immediately infer that the square of the characteristic determinant for each of the systems (L), (L^t) has the form

$$(\zeta_1^2 + \ldots + \zeta_n^2)^{2^{n-1}}$$

(consequently, in the determinant itself we must replace the exponent by 2^{n-2}).

Recall that we denote by $\omega_{\rm II}$ ($\omega_{\rm I}$) the collections of all forms of an even (odd) degree. Using the chains of orthogonal decompositions of the spaces $\mathbb{H}^r$ and the same arguments that we used in 2.2, 3.1, we get an analog of Theorem 1.

THEOREM 2. *For any $f_{\rm I}, f_{\rm II} \in \mathfrak{R}_\Delta$ there exists a unique solution of systems (L), (L^t) satisfying the conditions*

$$\omega_{\rm I}|_\Delta = \omega_{\rm II}|_\Delta = 0. \blacksquare \tag{6}$$

Theorem 2 and the considerations of 2.3 give an interesting statement that refers to the so-called *index* of the systems under consideration which is defined as the difference between the number of conditions imposed on the unknown functions that ensure the uniqueness of solution and the number of conditions imposed on the right-hand sides that ensure the solvability. In the ordinary notations we have

$$\operatorname{ind} L = \dim \operatorname{Ker} L - \dim \operatorname{Coker} L$$

(the index is frequently also defined as the difference between the number of linearly independent solutions of the given and the conjugate homogeneous system).

In our case, according to Theorem 2,

$$\operatorname{ind} L^t = \sum_{k=0}^{n} (-1)^k \dim \mathfrak{N}_\Delta^k = -\operatorname{ind} L,$$

where $\mathfrak{N}_\Delta^k$ is a subspace of $\mathfrak{N}_\Delta$ lying in $\mathbb{H}^k$.

As was pointed out in 2.3, $\dim \mathfrak{N}_\Delta^k = B_k$, i.e., it coincides with the corresponding Betti number for M. The alternated sum of Betti numbers is known as *Euler characteristic* of M (see [21, 31]). Hence the following statement.

PROPOSITION 2. *For systems of the type of the Cauchy–Riemann systems, considered on the manifold M, we have*

$$\operatorname{ind} L^t = \chi(M) = -\operatorname{ind} L.$$

Let us verify that the generalized solutions belong to the spaces W. In the general case, as in 2.2, the space $W \subset \mathbb{H}^r$ results from the completion of the linear manifold $\mathbb{H}^r \cap C^1$ with respect to the norm generated by the inner product

$$\{\omega, \chi\} = (d\omega, d\chi) + (\delta\omega, \delta\chi) + (\omega_\Delta, \chi_\Delta).$$

Considering the scalar squares on the right-hand and left-hand sides of each of the relations entering into (L) (into (L^t)) and using the fact that $(d\omega, \delta\omega) = 0$ for smooth forms, and the solutions under consideration are strong (admit of a smooth approximation), we obtain the estimate of the sum $(d\omega, d\omega) + (\delta\omega, \delta\omega)$ in terms of the right-hand sides for each of the forms entering into the system (and satisfying condition (6)). The collection of these estimates gives

$$|\omega_{\mathrm{I}}, W|^2 = (d\omega_{\mathrm{I}}, d\omega_{\mathrm{I}}) + (\delta\omega_{\mathrm{I}}, \delta\omega_{\mathrm{I}}) = |L\omega_{\mathrm{I}}, \mathbb{H}|^2 = |f_{\mathrm{II}}, \mathbb{H}|^2. \quad (7)$$

A similar estimate is obviously valid for the forms ω_{II} appearing in (L^t).

Relations (7) imply the belonging of the generalized solution to the space W and, at the same time, the proof of Theorem 2 independent of the use of orthogonal decompositions. Indeed, the uniqueness of the solution for (L) follows directly from (7) and the existence follows from the fact that the orthogonal complement of $\mathfrak{R}(L)$ in $\mathfrak{R}_\Delta$ belongs to $\mathfrak{N}(L^t)$, i.e., is empty (here we again use the equivalence of weak and strong definitions of the operators L, L^t).

The same reasoning is valid for (L^t).

EXAMPLE 1. The only n-dimensional compact manifold without boundary, that admits of the introduction of the Euclidean metric and is, therefore, ideally suitable for analyzing numerous models, is an n-dimensional torus T^n. The simplest way of obtaining it (precisely with the indicated metric) is the identification of opposite faces of an n-dimensional parallelepiped. When we deal with functional spaces, this corresponds to the consideration of functions subject to the corresponding periodicity conditions in the parallelepiped.

In particular, the space W appearing in our discussion is defined, in this case, as a completion of a linear manifold of smooth periodic functions in the corresponding metric. It immediately follows from this method of constructing W that every harmonic form on a torus is a system of constants. Using Hodge's theorem, we infer that Betti numbers for a torus coincide with the binomial coefficients, i.e., $B_p = C_p^n$ (they are equal to the dimension of the corresponding space of p-covectors). As a result, Theorem 2 yields the following statement.

PROPOSITION 3. *In the parallelepiped $V \subset \mathbb{R}^n$ for the systems (L), (L^t), for any collection of right-hand sides orthogonal to the constants, there exists a periodic solution, belonging to $W(V)$, defined with an accuracy to within 2^{n-1} arbitrary constants.*

COROLLARY. *On the indicated assumptions, we have*

$$\operatorname{ind} L = \operatorname{ind} L^t = 0.$$

EXAMPLE 2. It is easy to verify that for the n-dimensional sphere S^n we have $B_0 = 1$, $B_n = 1$, and the other Betti numbers are zeros. Thus $\chi(S^n) = 0$ for an odd n and $\chi(S^n) = 2$ for an even n, and this defines the index of the systems (L), (L^t) in this case.

3.3. Regular invariant systems; the spectrum. It is obvious that on the manifold of an arbitrary number of dimensions, the collection of invariant systems, i.e., systems that can be written by means of the operators d, δ, is far from being exhausted by the analogs of the Cauchy–Riemann equations which we have considered. In Sec. 6 we shall discuss some classical equations of mathematical physics which admit of this notation; and, for the time being, shall consider some "nonclassical" examples, resulting from the general discussion, restricting the examples to **regular** systems in which the number of unknown functions, the components of forms, is equal to the number of equations. Thus, underdetermined systems of the form $d\omega_{(r)} = f_{(r+1)}$, $n > 1$, $r > 0$, or overdetermined systems of the form $d\omega_{(r)} = f_{(r+1)}$, $\delta\omega_{(r)} = f_{(r-1)}$, $n > 2$, $r > 0$, and the like, whose study is, in general, of some interest, are not considered in this book.

Here is one of the techniques of obtaining regular systems. We take the system

$$d\omega_{(0)} - \omega_{(1)} = f_{(1)}, \qquad \delta\omega_{(1)} + \omega_{(0)} = f_{(0)} \tag{8}$$

for $n = 1$. Then: (a) this system is regular upon the transition from $n = 1$ to an arbitrary n; and (b) the systems

$$d\omega_{(p)} - \omega_{(p+1)} = f_{(p+1)}, \quad \delta\omega_{(p+1)} - \omega_{(p)} = f_{(p)}, \qquad p = 2, 3, \ldots$$

are regular for $n > 1$.

We can use a similar process proceeding from the system

$$\begin{aligned} d\omega_{(1)} + \omega_{(2)} &= f_{(2)}, \\ d\omega_{(0)} + \delta\omega_{(2)} - \omega_{(1)} &= f_{(1)}, \\ \delta\omega_{(1)} + \omega_{(0)} &= f_{(0)}, \end{aligned} \tag{9}$$

defined for $n = 2$.

We can present other similar constructions. The resulting systems possess a number of specific properties which are considered in [14, 51], for instance. Let us discuss one of them considering the following simplest example.

If we replace the minus sign by the plus sign in the first equation of (8), then the character of solvability of the system will prove to be dependent on the **metric** characteristics of the manifold M. This fact is equivalent to the statement on the existence of nontrivial solutions of the equations

$$d\omega_{(0)} + \lambda\omega_{(1)} = 0, \qquad \delta\omega_{(1)} + \lambda\omega_{(0)} = 0 \tag{10}$$

for certain λ (under the conditions ensuring the uniqueness of the solution for (8)), i.e., on the nonemptiness of the **point spectrum** of the corresponding operator.

A smooth solution of the homogeneous system (10) simultaneously satisfies the equations

$$-\Delta\omega_{(r)} - \lambda^2\omega_{(r)} = 0, \qquad r = 0, 1.$$

The spectrum of the Laplace operator on M (known as the "M spectrum"), which is one of the classical objects of investigation for the analysis on Riemannian manifolds [60], is not considered here.

At the same time, for the regular systems of the form (8), (9) given above, the corresponding theorems on the existence and uniqueness of generalized solutions are valid irrespective of the structure of M. Certain theorems on the smoothness of solutions are also valid. We can establish these facts by using techniques similar to those which were used in 3.2.

3.4. Splitting and the Lorentz metric. One of the interesting properties of the invariant systems under consideration, which is connected with their nature, is a specific splitting that can be carried out when the manifold M being considered is decomposed in the direct product $M = M_1 \times M_2$. In this case, the Hilbert space of the forms $\mathbb{H} = \oplus_0^n \mathbb{H}^k$ on M can be regarded as the sum of the corresponding tensor products

$$\mathbb{H}^k = \mathbb{H}_1^k \otimes \mathbb{H}_2^0 + \mathbb{H}_1^{k-1} \otimes \mathbb{H}_2^1 + \ldots + \mathbb{H}_1^0 \otimes \mathbb{H}_2^k, \tag{11}$$

where the subscript shows the number of the manifold over which the forms are considered. If $\dim M_1 < k$ (or $\dim M_2 < k$), then the corresponding terms in (11) are evidently absent. This splitting for invariant systems results from the use of decomposition (11) for forms, the formula

$$d(\omega \otimes \chi) = d\omega \otimes \chi + (-1)^r \omega \otimes d\chi \qquad (r \text{ — degree of } \omega)$$

and the corresponding formula for the operator δ.

We use this remark when $\dim M_1 = 1$. Denoting the collection of odd-degree forms entering into (L) by ω_1, we can write

$$\omega_{\rm I} = v_{(0)} \otimes u_{\rm I} + v_{(1)} \otimes u_{\rm II}, \tag{12}$$

where $u_{\rm I}$, $u_{\rm II}$ are collections of odd or even forms over M_2 whose dimension is assume to be equal to n.

REMARK. It should be remembered that the notation of the form (12) is of symbolic nature. For example, if $\dim M_2 = 2$, $\{\varphi_k(x)\}_0^\infty$ is a certain basis in the Hilbert space of covectors over M_2, $v_{(0)} = v_{(0)}(t)$, then

$$v_{(0)} \otimes u_{\rm I} = \sum_k v_k(t)\varphi_k(x),$$

i.e., the right-hand side contains a sum of functions of three variables.

Returning to (12), we can write

$$d\omega_{\rm I} = dv_{(0)} \otimes u_{\rm I} + v_{(0)} \otimes du_{\rm I} - v_{(1)} \otimes du_{\rm II},$$

$$\delta\omega_{\rm I} = v_{(0)} \otimes \delta u_{\rm I} + \delta v_{(1)} \otimes u_{\rm II} - v_{(1)} \otimes \delta u_{\rm II},$$

and the system of equations (L), which we represent symbolically as

$$(d+\delta)\omega_{\rm I} = f_{\rm II} \equiv f_{(0)} \otimes g_{\rm II} + f_{(1)} \otimes g_{\rm I},$$

will split into two subsystems

$$\begin{aligned} v_{(0)} \otimes (d+\delta)u_{\rm I} + \delta v_{(1)} \otimes u_{\rm II} &= f_{(0)} \otimes g_{\rm II}, \\ -v_{(1)} \otimes (d+\delta)u_{\rm II} + dv_{(0)} \otimes u_{\rm I} &= f_{(1)} \otimes g_{\rm I}. \end{aligned} \tag{13}$$

The meaning of splitting (13) becomes clearer when we make a certain change in the notations. We denote the first term on the right-hand side of (12) by φ and the second by ψ, and assume that M_1 is parametrized by the variable t by setting

$$d \equiv D_t \equiv -\delta$$

on M_1.

Then we can rewrite (13) in the form

$$\begin{aligned} -D_t\psi + (d+\delta)_n\varphi &= p, \\ D_t\varphi - (d+\delta)_n\psi &= q, \end{aligned} \tag{14}$$

where the operator $(d+\delta)_n$ acts with respect to the variables, namely, the coordinates on M_2.

Let us consider some examples. For $n = 1$ system (14) reduces to the inhomogeneous Cauchy–Riemann equations

$$-D_t\psi + D_x\varphi = p, \qquad D_t\varphi + D_x\psi = q.$$

For $n = 2$, using the notations

$$\varphi = (\varphi_1, \varphi_2), \qquad \psi = \psi_{(0)}, \psi_{(2)},$$

(in accordance with the numbering of the components u_{I}, u_{II} in (12)), we obtain, if we consider the metric on M_2 to be Euclidean, the representation

$$\begin{array}{ll} -D_t\psi_{(0)} - D_1\varphi_1 - D_2\varphi_2, & D_t\varphi_1 - D_1\psi_{(0)} - D_2\psi_{(2)}, \\ -D_t\psi_{(2)} + D_1\varphi_2 - D_2\varphi_1, & D_t\varphi_2 - D_2\psi_{(0)} + D_1\psi_{(2)}, \end{array} \tag{15}$$

for the left-hand sides of (14). In order to return from (15) to the componentwise form of the system

$$\delta\omega_{(1)} = f_{(0)}, \qquad d\omega_{(1)} + \delta\omega_{(3)} = f_{(2)}$$

(which we have split), for the chosen technique of ordering the components of the forms, it is convenient to regard t as the last (third) coordinate. Then $\varphi_1 = \omega_1$, $\varphi_2 = \omega_2$, $\psi_{(0)} = \omega_3$, $\varphi_{(2)} = \omega_{123}$ and (15) assumes the form

$$\begin{array}{ll} -D_3\omega_3 - D_1\omega_1 - D_2\omega_2 = f_{(0)}, & D_3\omega_1 - D_1\omega_3 - D_2\omega_{123} = f_{13}, \\ D_1\omega_2 - D_2\omega_1 - D_3\omega_{123} = f_{12}, & D_3\omega_2 - D_2\omega_3 + D_1\omega_{123} = f_{12}. \end{array}$$

It is also easy to formulate the general law of this (inverse) transition [14].

The described splitting turns out to be very convenient for proving, by induction on the dimensionality, various properties of forms or invariant systems [14]. We shall use splitting for considering the transition to the Lorentz metric on the manifold M of the type we have studied ($\dim M_1 = 1$) regarding t as "time", i.e., passing to a metric tensor, which, at every point, can be reduced to the form

$$g_{11} = \ldots = g_{nn} = 1, \qquad g_{n+1,n+1} = -1$$

(the other components are zeros).

In order to write system (14) in this case, it is sufficient to change the sign before the operator δ on M_1 since the definition of d does not depend on the metric; and when defining δ, we must change the sign in the inner product. This means that the term $-D_t\psi$ in (14) will be replaced by $D_t\psi$. The corresponding change must be performed in any other form of the notation of the equation.

For $n = 1$, after such a change the Cauchy–Riemann equations will pass into the simple hyperbolic system considered in 5.3, Ch. 1. For $n = 2$, we obtain the so-called **symmetric hyperbolic system** [62], which we shall discuss in greater detail in 4.4.

After the indicated change of the metric tensor, the operator $d\delta + \delta d$ on M will split into a chain of wave operators applied to every component of the form. The characteristic determinant of the systems (L), (L^t) assumes the form

$$(\zeta_1^2 + \ldots + \zeta_n^2 - \zeta_{n+1}^2)^{2^{n-1}}.$$

Similar constructions can also be carried out in the case of metric tensors of an arbitrary signature, but we shall not consider them here.

4. Boundary Value Problems for Invariant Systems

4.0. Preliminary remarks. As we have repeatedly emphasized, one of the most important examples of Riemannian manifold, the atlas of which consists of a single map and the metric tensor has the form defined by (12), Sec. 3, Ch. 1, is the domain V of the Euclidean space $\mathbb{R}^n$. The closure $\overline{V}$ of such a domain is an example of a manifold with boundary. In this section we shall deal precisely with this kind of manifold.

The definition of the domain $V \subset \mathbb{R}^n$ we have formulated makes us employ, in a Euclidean space as well, only objects that have an invariant meaning and use techniques that admit of a generalization to an arbitrary Riemannian manifold. Correspondingly, we again deal with forms, their integration, with the operators d, δ, invariant systems, etc.

There is a standard technique ("glueing of the double") that turns a manifold with boundary (sufficiently regular) into a manifold without boundary. This technique will be considered in 4.1. The transition to a manifold without boundary makes it possible to overcome the difficulties connected with the problem of the equivalence of weak and strong definitions of the operators d, δ, L, L^t, etc. This transition also allows us to formulate special classes of boundary conditions for the forms under consideration (arising as the conditions of the "continuability to the double" with the preservation of the smoothness) which are now included into the description of the domain of definition of differential operations. The theorems on the existence and uniqueness of solutions of the equations considered in Secs. 2, 3 give the corresponding results for the solutions of boundary value problems.

Note that an alternative approach to the study of boundary value problems for invariant systems is the consideration of their difference analogs (Ch. 3).

4.1. Glueing of the double. Let us discuss in greater detail the operation of completion of the domain $V' \subset \mathbb{R}^n$ with a smooth boundary, which is homeomorphic to a ball, to the smooth manifold M, which is homeomorphic to an n-dimensional sphere. A similar construction can be carried out in a considerably more general situation, i.e., is applicable to an arbitrary smooth manifold with boundary. In the sequel, we shall use some generalizations that refer to the domains $V \subset \mathbb{R}^n$ of a more complicated structure. However, for the time being, we shall restrict our discussion to the indicated simple case which is convenient for the analysis of the problems of interest to us.

Together with the domain $V' \subset \mathbb{R}^n$ referred to the coordinates $x^1, \ldots, x^n$ we introduce its counterpart $V'' \subset \mathbb{R}^n$. We call the points V', V'', whose coordinates coincide, *respective points.* Considering now V', V'' to be two distinct domains and identifying the respective points of the boundary, we get the n-dimensional manifold $M = \overline{V}' \cup \overline{V}''$ which is homeomorphic to the n-dimensional sphere. We shall show that on the assumption of smoothness of the boundary $S = \partial V' = \partial V''$ on M, we can introduce a differential structure (4.2, Ch. 1).

Alongside the global coordinate system (x) in V', we introduce a similar system (y) in V'', assuming that

$$y^1 = -x^1, \qquad y^i = x^i, \quad i = 2, \ldots, n$$

for the respective points. Since we can take the systems (x), (y) as local coordinates in the open domains V', V'', in order to prove the existence of a smooth atlas meeting the requirements set up in 4.2, Ch. 1, it is sufficient to consider the cover of the *edge*, namely, the identified points of the boundaries of V', V''.

REMARK. Note that the **orientability** of M, necessary for using the results of Secs. 2, 3, is established precisely by the indicated correspondence between the local coordinates of the two principal maps. (It stands to reason that it admits of many versions.) In particular, this correspondence yields the relations

$$\partial M = 0, \qquad \int_M \omega_{(n)} = \Big(\int_{V'} + \int_{V''} \Big) \omega_{(n)}.$$

Suppose now that the neighborhood of the boundary of V' in $\mathbb{R}^n$ is covered by a finite number of domains V'_σ such that

$$\varphi'_\sigma : U'_\sigma \to V'_\sigma, \qquad x^i = \varphi^i_\sigma(\xi), \quad i = 1, \ldots, n, \tag{1}$$

gives a smooth map with a positive Jacobian of the half-ball

$$\Big\{ \xi : \sum_1^n (\xi^i)^2 < 1,\ \xi^1 \leq 0 \Big\}$$

of the coordinate space (ξ) on $V'_\sigma \cap \overline{V}'$. In this case, we regard $x^1, \ldots, x^n$ as local coordinates in $V'_\sigma \cap V'$ (defined by the coordinate system (x) in V') and assume that the equation of the part of the boundary S lying in V'_σ has the form

$$\xi^1 = 0. \tag{2}$$

Obviously, the smoothness of the functions φ^i_σ that define φ'_σ is limited only by the smoothness of the boundary S.

Suppose that V''_σ is a set of points corresponding to the points V'_σ. The collection of the domains

$$V_\sigma = (V'_\sigma \cap \overline{V}') \cup (V''_\sigma \cap \overline{V}'')$$

lying in M gives an open cover for the edge of M. If $y^1, \ldots, y^n$ are local coordinates in $V''_\sigma \cap V''$, defined by the coordinate system (y) in V'', then the system of functions

$$y^1 = -\varphi'_\sigma(-\xi^1, \xi^2, \ldots, \xi^n),$$

$$y^i = \varphi^i_\sigma(-\xi^1, \xi^2, \ldots, \xi^n), \quad i = 2, \ldots, n, \tag{3}$$

gives a smooth invertible mapping with a positive Jacobian

$$\varphi''_\sigma : U''_\sigma \to V''_\sigma,$$

that maps the half-ball

$$\left\{\xi : \sum_1^n (\xi^i)^2 < 1,\ \xi^1 \geq 0\right\} \quad \text{on} \quad V'_\sigma \cap \overline{V}''.$$

Regarding now the restriction of the map

$$\varphi_\sigma : (U'_\sigma \cup U''_\sigma) \to (V'_\sigma \cup V''_\sigma)$$

(coincident with φ'_σ, φ''_σ on the corresponding domains) to the unit ball $\sum_1^n (\xi^i)^2 < 1$ as a restriction defining the local coordinates of (ξ) in V_σ, we see that these coordinates are smoothly connected with the local coordinates (x), (y) in $V_\sigma \cap V'$, $V_\sigma \cap V''$. The smoothness of the coordinate transformations at the intersections $V_\sigma \cap V_\tau$ of the open sets that cover the edge and the positivity of the Jacobians are ensured automatically. In this case, the edge is a submanifold of M in the sense of the definition from 4.2, Ch. 1.

Let us now consider the following important circumstance. As on any smooth manifold, we can introduce on the manifold M, which we have constructed, a Riemannian metric which is also smooth. However, we must weaken this smoothness if we require that the metric should coincide in V', V'' with the Euclidean metric which is induced by the enveloping space $\mathbb{R}^n$. We shall accept precisely this requirement since it is convenient for our purposes and consider its consequences.

Thus we want to turn M into a Riemannian manifold, supposing that the metric in V', V'' (in the coordinates (x), (y)) is Euclidean. We shall show that this assumption induces the metric on M, i.e., uniquely defines on M the metric tensor which turns out to be Lipschitz-continuous on the edge. Again, we shall consider only the neighborhood of the edge.

Since we suppose the boundary of V' to be smooth and can take its cover arbitrarily fine, we can assume that the coordinate ξ^1 appearing in (2) is the distance to the hyperplane (2) along the normal. Then, in the coordinates (ξ) (we omit the indication of the number of the neighborhood), the components of the metric tensor satisfy the relation

$$g_{1'j'} = 0, \qquad j = 2, \dots, n.$$

In accordance with the tensor law of transformation, the other components can be found from the relations

$$g_{i'j'} = D_{i'}\varphi^i D_{j'}\varphi^j g_{ij}$$

(summation over the repeating indices). It follows from relations (1), (3) that in this case the components $g_{i'j'}(\xi)$ are functions, which are even relative to hyperplane (2), and, consequently, Lipschitz-continuous in the neighborhood of the edge. In general, the first derivatives of the metric tensor suffer a jump on the edge.

Hence follows a remark which is of importance to us. When considering the forms and the operators d, δ on M, we must take into account that the definition of δ depends on the metric tensor. The expression for $\delta\omega$ in local coordinates in the neighborhood of the edge contains derivatives along these coordinates of the products of the components of the metric tensor by one or another component of the form ω. Thus the coefficients in the derivatives of the components of ω are Lipschitz-continuous and the coefficients in the subordinate terms can have discontinuities of the type of a jump (on the edge).

4.2. Invariant systems in bounded domains. When considering, in a finite domain, one of the principal invariant objects discussed in the preceding sections, namely, the Laplace operator, there is no need to carry out constructions as we did in 4.1. On the assumption that the required functions, the components of the forms, are subject to the Dirichlet or Neumann conditions on the boundary of the domain V', we can study the equation $-\Delta\omega = f$ with the use of standard methods (cf. 5.3, Ch. 1), and the specificity of the viewpoint formulated in Sec. 0 does not manifest itself.

The situation is quite different as concerns first-order invariant systems. Here the boundary value problems have been thoroughly investigated only for $n \leq 2$, and the construction carried out in 4.1 proves to be very useful for obtaining general results (referring both to boundary value problems and to orthogonal decompositions considered in 5.2).

Preserving the assumptions and notations of 4.1, we shall speak of the systems (L), (L^t) considered in 3.2. It is natural to use the following scheme. We set a one-to-one correspondence between the system of equations $L\omega'_{\mathrm{I}} = f'_{\mathrm{II}}$ in V' and a certain system in M

whose solution ω_{I} a priori exists and is unique. The restriction of ω_{I} to V' must give the unique solution of the original problem. The simplest variant of the realization of this scheme that suggests itself is the introduction of the form $\omega''_{\mathrm{I}} = \omega_{\mathrm{I}}|_{V''}$ such that $\omega''_{\mathrm{I}} = \omega'_{\mathrm{I}}$ at the corresponding points. However, it turns out that this straightforward approach needs a modification since otherwise the components of ω_{I} will suffer a discontinuity on the edge.

As would be expected, the necessary modification leads to a special class of homogeneous boundary conditions for ω'_{I} (or ω'_{II}) which ensure the existence and uniqueness of the solution of the equations (L) (or (L^t)). The form of these conditions depends on the choice of the coordinate systems (x), (y), on the equations of the boundary of V', and on the local coordinates introduced in (2), (3).

It should be emphasized that when considering (L), (L^t) in V', writing the Green formula

$$(L\omega_{\mathrm{I}}, \chi_{\mathrm{II}}) = \int\limits_{\partial V'} \mathcal{L}[\omega_{\mathrm{I}}, \chi_{\mathrm{II}}] + (\omega_{\mathrm{I}}, L^t\chi_{\mathrm{II}})$$

for the forms $\omega_{\mathrm{I}}, \chi_{\mathrm{II}} \in C^1(\overline{V}')$, and analyzing the structure of the integral of $\partial V'$, we can find homogeneous (dual) boundary conditions for ω_{I}, χ_{II} (similar, on the whole, to the conditions given below) which ensure the validity of the inequalities (cf. (7), Sec. 3)

$$|\omega_{\mathrm{I}}, W| \le C|L\omega_{\mathrm{I}}, \mathbb{H}|, \qquad |\chi_{\mathrm{II}}, W| \le C|L^t\chi_{\mathrm{II}}, W|, \tag{4}$$

without carrying out constructions from 4.1. However, precisely the latter constructions allow us to overcome a considerable technical difficulty connected with the proof of the **equivalence** of weak and strong definitions of the operators L, L^t. Without this equivalence, inequalities (4) can give only the uniqueness of the strong solution, leaving the question concerning its existence for any $f \in \mathbb{H}$ open (cf. 5.3, Ch. 1).

Thus suppose that we have constructed $M = \overline{V}' \cup \overline{V}''$ in accordance with the constructions from 4.1 and ω' is a continuous form of degree p defined on V' in coordinates (x). We agree to label the system of subscripts $i_1 \dots i_p$, where i_k $(k = 1, \dots, p)$ can assume the values $1, \dots, n$, by one letter i. We divide all these systems of subscripts into two groups, one group containing the subscript 1 and the other not containing it. This division is connected with the existence of the "preferred coordinate" x^1 (or y^1) in (x) (or (y)). We associate ω' with two types of forms ω'' on V'' supposing that

$$\begin{aligned} \omega_i''(y) &= \omega_i'(x), \quad 1 \notin i, \\ \omega_i''(y) &= -\omega_i'(x), \quad 1 \in i, \end{aligned} \tag{5}$$

or

$$\begin{aligned} \omega_i''(y) &= -\omega_i'(x), \quad 1 \notin i, \\ \omega_i''(y) &= \omega_i'(x), \quad 1 \in i, \end{aligned}$$

at the corresponding points. The form ω (coincident with ω' on V' and with ω'' on V'') defined on $V' \cup V''$ in this way is said to be *even* in the first case and *odd* in the second case. Let us consider, for definiteness, the **even** forms, having found out under what conditions they will preserve the continuity on the edge.

Using relations (1), (3) (omitting the subscript σ), we have, in accordance with the tensor law of transformation of forms (cf. examples from 4.3 below)

$$\begin{aligned} \omega_{i'}(\xi) &= D_{i'}\varphi^i(\xi)\omega_i(x(\xi)), \quad \xi^1 < 0, \\ \omega_{i'}(\xi) &= D_{i'}\varphi^i(\xi)\omega_i(y(\xi)), \quad \xi^1 > 0, \end{aligned}$$

where i, i' are the corresponding systems of subscripts, and the summation is carried out over the repeating subscripts. Let us find the conditions of continuity of $\omega_{i'}(\xi)$. According to (5), on the superplane $\xi^1 = 0$ we have

$$\omega_i'' = \omega_i', \quad 1 \notin i, \qquad \omega_i'' = -\omega_i', \quad 1 \in i.$$

In addition, $D_{i'}\varphi^i$ changes sign for $1 \notin i'$, $1 \in i$, or for $1 \in i'$, $1 \notin i$, and does not change sign in other cases. It follows that the validity of the relations (on the edge)

$$\omega_{i'} = D_{i'}\varphi^i(\xi)\omega_i(x(\xi)) = 0, \qquad 1 \in i' \tag{6}$$

is the necessary and sufficient condition for the continuity of an even form in the neighborhood of the edge.

Similar reasoning shows that for odd forms the conditions of continuity are given by the relations

$$\omega_{i'}(\xi)|_{\xi^1=0} = 0, \qquad 1 \notin i'. \tag{7}$$

Note that it follows from the invertibility of mappings (1), (3) that we can write conditions (7), (6) in the coordinates (x).

Passing to the equations (L), (L^t), we note, first of all, that the definition of the space W given in 2.2 preserves sense for the manifold M we have constructed, i.e., the Lipschitz continuity of the metric tensor in the neighborhood of the edge is quite sufficient. In addition, since the structure of M is simple (it was obtained by "doubling" the domain of the Euclidean space and is homeomorphic to the ball), inequality (13), Sec. 2, is on M a consequence of the simplest classical inclusion theorem, and the fact that the basis of the subspace $\mathfrak{N}_\Delta$ consists of a pair of forms $\omega_{(0)}$, $\omega_{(n)}$ defined by constants. For forms belonging to $W(V')$ (to the restriction of $W(M)$ to V'), conditions of the form (6), (7) retain sense; and this allows us to formulate the following statement.

PROPOSITION 1. *The forms on M obtained by an even (odd) extension of the forms from $W(V')$, subject to conditions* (6) (*conditions* (7)), *belong to $W(M)$.*

PROPOSITION 2. *The forms from $W(M)$, which are even (odd), satisfy conditions* (6) (*conditions* (7)).

These statements are obvious enough. Their pedantic verification is given in [14]. ■

We assume the *generalized solution* of systems (L), (L^t) in V', that satisfy conditions (6) or (7), to be the forms $\omega_{\rm I}$, $\omega_{\rm II}$ from $W(V')$ which satisfy (in the corresponding sense) the equations and boundary conditions.

We shall formulate the fundamental theorem in the case of system (L) and conditions (6).

THEOREM 1. *The generalized solution of the equations (L) in V' under conditions* (6) *exists and is unique for any right-hand side of $f_{\rm II} \in \mathbb{H}(V')$ satisfying the condition*

$$(f_{(0)}, 1)_{V'} = 0. \tag{8}$$

PROOF. Completing V' to M and continuing the right-hand sides evenly to M, we can ascertain the existence on M of the solution $\omega_{\rm I}$ of the equation (L) (the condition $f_{\rm II} \in \mathfrak{R}_\Delta$ that reduces, in addition to the requirement $(f_{(0)}, 1)_M = 0$, to the requirement

$$\int\limits_M f_{(n)} = 0 \tag{9}$$

for even n, is automatically satisfied by virtue of the evenness of $f_{\rm II}$). This solution is always even (since the application of the operator

L does not violate the evenness). It belongs to $W(M)$ and, consequently, satisfies condition (6). It is unique on M (the condition $\omega_{\mathrm{I}}|_{\Delta} = 0$ reduces to a requirement of form (9) on $\omega_{(n)}$ for odd n and is again satisfied by virtue of the evenness of ω_{I}). It satisfies the system (L) in V' and is unique in V'. Indeed, if there existed another solution belonging to $W(V')$ and satisfying (6), then, being continued evenly, it would give a different solution of the system (L) on M. ■

A similar theorem is valid for (L) under conditions (7) and for (L^t) under conditions (6) or (7). We have only to make the requisite changes in the conditions of solvability and uniqueness.

The constructions we have used are, in essence, the generalization of the classical symmetry principle that reduces, in the simplest form, to the following. If

$$D_1 u_1 + D_2 u_2 = 0, \qquad D_1 u_2 - D_2 u_1 = 0, \qquad D_k = \frac{\partial}{\partial \xi^k}, \tag{10}$$

are homogeneous Cauchy–Riemann equations, then the function $u = (u_1, u_2)$, that satisfies (10) in the half-plane $\xi^1 > 0$, satisfies (10) for $\xi^1 < 0$ as well if we set

$$u_1(\xi^1, \xi^2) = -u_1(-\xi^1, \xi^2), \qquad u_2(\xi^1, \xi^2) = u_2(-\xi^1, \xi^2).$$

In order to ensure the continuity (and, hence, analyticity) of u for $\xi^1 = 0$, it is sufficient to require that $u_1(0, \xi^2) = 0$.

This approach can, obviously, also be used to obtain the theorems of the existence and uniqueness of solutions of boundary value problems for regular systems described in 3.3. Peculiar modifications appear in "splitting" domains considered in 3.4 (cf. [14]).

4.3. A disk, a ball, and a cube. I shall illustrate the content of 4.1 and 4.2 by some simple examples.

3A. A disk. Let $V' : (x^1)^2 + (x^2)^2 < 1$, $V'' : (y^1)^2 + (y^2)^2 < 1$. The form of the systems (L), (L^t) was given in 3.1. The corresponding points in V', V'' are

$$y^1 = -x^1, \qquad y^2 = x^2.$$

Covering the edge by neighborhoods and rectifying it, we shall actually use polar coordinates. We preserve the ordinary notations, setting $\xi^1 = r$, $\xi^2 = \vartheta$. We use the functions

$$x^1 = (1+r)\cos\vartheta, \qquad x^2 = (1+r)\sin\vartheta \tag{11}$$

to map the rectangle

$$-\frac{1}{2} < r \le 0, \qquad -\frac{2}{3}\pi < \vartheta < \frac{2}{3}\pi \tag{12}$$

onto the neighborhood U_1' of the boundary of V'. The same relations (11) give the map of the rectangle

$$-\frac{1}{2} < r \le 0, \qquad \frac{1}{3}\pi < \vartheta < \frac{5}{3}\pi \tag{13}$$

on the neighborhood U_2'. The pair of these neighborhoods completely covers the neighborhood of the boundary. The corresponding cover for V'' is given by the mapping

$$y^1 = -(1-r)\cos\vartheta, \qquad y^2 = (1-r)\sin\vartheta, \tag{14}$$

where $0 \le r < 1/2$, and ϑ varies in the same range (12), (13). In order to write the transformation laws, we need the derivatives $D_{i'}\varphi^j$. For the neighborhoods U_1', U_2' we have

$$\begin{aligned} D_{1'}\varphi^1 &= \cos\vartheta, & D_{2'}\varphi^1 &= -(1+r)\sin\vartheta, \\ D_{1'}\varphi^2 &= \sin\vartheta, & D_{2'}\varphi^2 &= (1+r)\cos\vartheta. \end{aligned} \tag{15}$$

Taking into account that the metric tensor has the form $g_{11} = g_{22} = 1$, $g_{12} = 0$ in the coordinates (x), we obtain

$$g_{1'1'} = 1, \qquad g_{1'2'} = 0, \qquad g_{2'2'} = (1+r)^2. \tag{16}$$

For the neighborhoods U_1'', U_2'', only the component $g_{2'2'} = (1-r)^2$ is different from (16). This component suffers a jump of the derivative for $r = 0$.

Let us consider the forms. We write the transformation relations for U_1', U_2'. The scalar $\omega_{(0)}$ is an invariant, for $\omega_{(1)}$ we have $\omega_{i'} = D_{i'}\varphi^j\omega_i$ or

$$\omega_{1'} = \omega_1\cos\vartheta + \omega_2\sin\vartheta, \qquad \omega_{2'} = (1+r)(-\omega_1\sin\vartheta + \omega_2\cos\vartheta).$$

Taking into account the skew-symmetry of $\omega_{21} = -\omega_{12}$, we get for $\omega_{(2)}$ the relation

$$\omega_{1'2'} = (D_{1'}\varphi^1 D_{2'}\varphi^2 + D_{1'}\varphi^2 D_{2'}\varphi^1)\omega_{12} = (1+r)\omega_{12}.$$

Considering the type of the boundary conditions that ensure, in accordance with the results of the preceding subsection, the existence and uniqueness of the generalized solutions of the systems (L), (L^t), we shall discuss the case of even forms. Then, according to (5), we must have the relation

$$\omega'_{(0)} = \omega''_{(0)}, \quad \omega'_1 = -\omega''_1, \quad \omega'_2 = \omega''_2, \quad \omega'_{12} = -\omega''_{12}$$

at the corresponding points, and, according to (6), the conditions

$$\omega_1 \cos\vartheta + \omega_2 \sin\vartheta = 0, \qquad \omega_{12} = 0 \tag{17}$$

must be satisfied on the boundary of the disk. Under the first condition (for the system (L)), the right-hand side must satisfy condition (8). Under the second condition (for the system (L^t)), $\omega_{(0)}$ will be defined with an accuracy to within a constant.

3B. A ball. Let $V' : (x^1)^2 + (x^2)^2 + (x^3)^2 < 1$. We shall define V'' and the coordinates of the corresponding points in the same way as we did in example 3A. The analogs of maps (11) are

$$\begin{aligned} x^1 &= (1+r)\cos\vartheta\cos\psi, \\ x^2 &= (1+r)\cos\vartheta\sin\psi, \\ x^3 &= (1+r)\sin\vartheta. \end{aligned} \tag{18}$$

For the coordinates (y), $1+r$ is replaced by $1-r$ and the map for y^1 is taken with the minus sign (cf. (14)). The ranges of variation of the parameters, for which the covering of the neighborhood of the edge is ensured, can be easily indicated. Considering r to be the first coordinate, ϑ to be the second coordinate, and ψ to be the third coordinate, we can write the matrix $\|D_{i'}\varphi^j\|$ and the expressions for $\omega_{i'}$, $\omega_{i'j'}$, $\omega_{1'2'3'}$ proceeding from (18). In the case of even forms, by adding to (L) (to (L^t)) the boundary conditions given (for $r = 0$) by the relations $\omega_{1'} = \omega_{1'2'3'} = 0$ (or $\omega_{1'2'} = \omega_{1'3'} = 0$), we shall obtain the problems which are of interest to us.

REMARK. For the system (L), the boundary conditions imposed on the covector $\omega_{(1)}$ can always be reduced to the classical form of conditions imposed on a "normal component". Suppose that the equation for the boundary can be solved, in the neighborhood under consideration, for $x^1 = \varphi(x^2, \dots, x^n)$, and that $(\alpha_1, \dots, \alpha_n)$ is a unit normal vector

$$\alpha_1 = \mu, \qquad \alpha_k = -\mu D_k\varphi, \quad k = 2, \dots, n,$$

$$\mu^2 = [1 + (D_2\varphi)^2 + \ldots + (D_n\varphi)^2].$$

We introduce the coordinates (ξ), setting

$$x^1 = \xi^1\alpha^1 + \varphi(\xi^2, \ldots, \xi^n), \qquad x^k = \xi^1\alpha_k + \xi^k, \quad k = 2, \ldots, n,$$

where $\alpha_i = \alpha_i(\xi^2, \ldots, \xi^n)$. Then

$$\omega_{1'} = D_{1'}\varphi^i\omega_i = \sum_1^n \omega_i \cos nx^i$$

and the condition $\omega_{1'} = 0$ is the condition for the vanishing of the normal component of the covector $\omega_{(1)}$.

Let us note, in passing, that only for $\omega_{(0)}$ (and for $\omega_{(n)}$ in some cases) is the form of the boundary conditions independent of the form of the boundary (of the equations defining it).

3C. A cube. In a number of cases, the constructions we have used admit of a modification that makes it possible to describe the boundary value problems for invariant systems in domain with piecewise-smooth boundary. We shall discuss the case, where V' is a cube of the Euclidean space,

$$V' : |x^i| < 1, \qquad i = 1, \ldots, n.$$

We shall use the same scheme as we used above to define the cube V'' and the manifold $M = \overline{V}' \cup \overline{V}''$, which is no longer smooth (say, in the neighborhood of the point $(1, \ldots, 1)$).

The smoothness of M was used above in two cases, namely, when we verified the equivalence of weak and strong definitions of the operators d, δ and when we defined the space $W(M)$. Nevertheless, it turns out to be possible to extend the results of 4.2 to the case of a nonsmooth M of a special form, that we have introduced, by using some auxiliary constructions.

Suppose that $P : \ |\xi^i| \leq 3$, $i = 1, \ldots, n$, is a cube of the Euclidean space referred to the coordinates (ξ). We divide P by the hyperplanes

$$\xi^i = 1, \quad \xi^i = -1, \qquad i = 1, \ldots, n \tag{19}$$

into 3^n equal cubes. Every one of these P_σ cubes can be obtained by a finite number of successive mappings in hyperplanes (19) from the central one: $|\xi^i| \leq 1$, $i = 1, \ldots, n$. In this case, the evenness or oddness of the number of mappings having been used does not depend on the order in which they were carried out. We identify the cubes P_σ obtained by means of an even number of mappings with V' and those obtained by an odd number of mappings with V''. Introducing, in every one of P_σ, the local coordinates (ξ_σ), obtained by the translation of the basic system (ξ) to the center of P_σ and establishing, in a proper way, the relationship between (ξ_σ) and (x) or (y), we can identify the even (or odd) forms on M (subject to boundary conditions that ensure the continuity) with the corresponding forms in P, considered in the neighborhood $|\xi^i| < 3/2$ of the central cube. At the same time, we identify the neighborhood of the boundary of the central cube with that of the edge of M, and this allows us to overcome the difficulties indicated above. Thus we obtain the following statement.

PROPOSITION 3. *Theorem* 1 *and its analogs remain valid in the domain* V' *that we have considered.* ■

Let us elucidate the explicit form of the boundary conditions added to the system (L^t) in the cube $V' \subset \mathbb{R}^3$ when the indicated scheme is used. Suppose that S^k is a pair of faces $x^k = \pm 1$, $k = 1, 2, 3$. When considering even forms, we obtain

$$\omega_{12}|_{S^1} = \omega_{13}|_{S^1} = \omega_{12}|_{S^2} = \omega_{23}|_{S^2} = \omega_{13}|_{S^3} = \omega_{23}|_{S^3} = 0,$$

and for $\omega_{(0)}$ we get the additional condition $(\omega_{(0)}, 1) = 0$.

For odd forms we have

$$\omega_{(0)}|_{S^1 \cup S^2 \cup S^3} = 0, \qquad \omega_{23}|_{S^1} = \omega_{13}|_{S^2} = \omega_{12}|_{S^3} = 0,$$

and for the right-hand sides, the solvability condition $(f_{(3)}, 1) = 0$.

If we use the possibility of interpreting (in $\mathbb{R}^3$) the form $\omega_{(2)}$ as a vector, i.e., $\omega_{(2)} = *v_{(1)}$ $(v_1 = \omega_{23},\ v_2 = -\omega_{13},\ v_3 = \omega_{12})$, then for the conditions we have found we can indicate a "hydrodynamic analog": either all tangent components of the "velocity" $v_{(1)}$ are defined on the boundary or the normal component and the "pressure" $\omega_{(0)}$.

REMARK. This scheme which uses mappings can also be applied, as it is easy to note, to a number of other types of manifolds M with an irregular boundary.

4.4. Time-dependent systems.

Let us consider briefly correct boundary value problems for systems with a singled out variable, namely, the time, resulting from (L), (L^t) after we pass to the

Lorentz metric described in 3.4. These systems belong to the class of the so-called **symmetric hyperbolic** systems [12, 62].

The hyperbolicity of this type of system has already been noted and the symmetry (in our case the existence of the equality $A^t = -A$, where A is the corresponding operator) follows immediately from (14), Sec. 3, since the operators $(d+\delta)_n$ acting onto φ and onto ψ (the collection of even and odd forms) are formally conjugate.

REMARK. In the case of two space variables, it immediately follows from (15), Sec. 3, that upon the ordering

$$(\psi_{(0)}, \psi_{(2)}, \varphi_1, \varphi_2) = w$$

we have, in matrix form,

$$Aw \equiv (A^0 D_t + A^1 D_1 + A^2 D_2)w,$$

$$A^1 = \begin{pmatrix} 0 & 0 & -1 & 0 \\ 0 & 0 & 0 & 1 \\ -1 & 0 & 0 & 0 \\ 0 & 1 & 0 & 0 \end{pmatrix}, \qquad A^2 = \begin{pmatrix} & & & -1 \\ 0 & & -1 & \\ & -1 & 0 & \\ & -1 & & \end{pmatrix}, \tag{20}$$

and the matrix A^0 (after the change of sign before $D_i\psi$) is an identity matrix.

(a) **The Cauchy problem in the domain of the Euclidean space.** Suppose that the domain $Q \subset \mathbb{R}^{n+1}$ is bounded by two cones

$$(x^1)^2 + \ldots + (x^n)^2 = (x^{n+1} \pm 1)^2, \qquad |x^{n+1}| \le 1,$$

and $V \subset Q$ is a "lense-shaped" domain that lies inside Q, i.e., is bounded by two space-like surfaces S_+, S_- such that $\cos \widehat{nx}^{n+1}$ is positive on S_+ and negative on S_-. Suppose that $\mathbb{H}(Q)$ is a Hilbert space of the system of functions $w = (w_1, \ldots, w_{2^{n+1}})$ (we number the components of the forms as it is done in the remark) and $A : \mathbb{H} \to \mathbb{H}$ is the closure in $\mathbb{H}$ of the hyperbolic operator, corresponding to system (14), Sec. 3, which was initially defined in Q on smooth functions subject to the **Cauchy conditions**

$$w|_{S_-} = 0. \tag{21}$$

The operator $A^t : \mathbb{H} \to \mathbb{H}$ can be defined by analogy, with the replacement of (21) by the conditions imposed on S_+. Papers [12, 14] give the following theorem.

THEOREM. *For any $f \in \mathbb{H}$ there exists a unique solution of the Cauchy problem* (*strong,* cf. 5.3, Ch. 1) *for the equation*

$$Aw = f, \tag{22}$$

which satisfies the inequality

$$|w, \mathbb{H}| \leq C|f, \mathbb{H}|. \blacksquare$$

The corresponding theorem is also valid for A^t.

(b) **The Cauchy problem in $M = M_1 \times M_2$.** It is supposed that M_1 is the interval $[0, b]$ of the real axis and M_2 is a smooth n-dimensional manifold without boundary. The definition of the operators $A, A^t : \mathbb{H}(M) \to \mathbb{H}(M)$ is similar to that given above with the only difference that the part of the surfaces $S_\pm$ is played by the boundary surfaces $b \times M_2$, $0 \times M_2$. The proof of the theorem on the existence and uniqueness of the solution of the corresponding Cauchy problem, based on the methods used in the theory of symmetric systems, is given in [14]. Of course, a different approach is also possible which is used in the study of operator equations of the form

$$D_t w - Fw = f$$

with the elliptic operator F.

(c) **Mixed problems in the Euclidean cylinder.** In this case the domain under consideration is again the direct product $M_1 \times M_2$, where M_1 is the same interval and M_2 is a bounded domain of the Euclidean space with a smooth boundary. When formulating a problem for Eq. (22), in addition to the Cauchy conditions we must introduce boundary conditions on the lateral surface $[0, b] \times \partial M_2$. They are defined in accordance with the constructions carried out in 4.2. Then, the glueing of the double reduces the mixed problem to the Cauchy problem of the case (b).

(d) **The Goursat problem.** The specific character of the systems being considered is especially vividly manifested in the possibility of posing the Goursat problem for them. In this case, the domain under consideration is the domain Q from (a). The operator A is defined as the closure in $\mathbb{H}(Q)$ of the same operator, corresponding to (14), but initially defined on smooth functions subject, on the lower boundary of the characteristic cone, to linear relations 2^{n-1} in number defined by the structure of the matrices A^k (see (20)) and the geometry of the cone. By means of similar conditions, on the upper boundary we can define the operator A^t. The corresponding arguments are given in [14, Ch. 3, Sec. 4] together with the following theorem.

THEOREM. *For any $f \in \mathbb{H}$, the solution of the Goursat problem for Eq.* (22) *exists and is unique. The corresponding result is also valid for the operator A^t.*

EXAMPLE. If we introduce polar coordinates in the plane $x^3 \equiv t = 0$ for $n = 2$ denoting the angle by ϑ, then the boundary conditions on the lower part of the cone for the hyperbolic system obtained from (15) by the substitution of $D_t\psi$ for $-D_t\psi$ assume the form

$$\psi_{(0)} \cos\vartheta - \psi_{(2)} \sin\vartheta + \varphi_1 = 0,$$

$$\psi_{(0)} \sin\vartheta - \psi_{(2)} \cos\vartheta + \varphi_2 = 0.$$

The constructions carried out in this subsection can be extended, in a natural way, to systems that originate from those given in 3.3 when we passed to the Lorentz metric [14]. It would be of interest to consider in this context the problem of alteration of the type ("mixed type problems").

5. Special Constructions on Manifolds with Boundaries

5.1. Index in boundary value problems. In 3.2 we introduced the concept of an index of a system of equations and found out that for the multidimensional Cauchy–Riemann system, considered on a smooth manifold without boundary, the index coincided with the Euler characteristic. When we pass to a manifold with boundary and boundary value problems, the definition of the index (this time the index of a boundary value problem) does not need any modification, but the arguments that allow us to calculate it are, as a rule, more complicated.

For the class of problems introduced in Sec. 4, the complications that arise are not very essential. Actually, the result being established coincides with that established in 3.2. We shall formulate it considering V' to be an arbitrary smooth manifold with boundary.

As was already noted, all the constructions carried out in Sec. 4 (referring, in the main, to the case of a sphere) automatically preserve their sense in this, more general, situation.

THEOREM 1. *The index of the systems* (L), (L^t), *considered on the manifold* V' *with the boundary conditions (and the definition of the solution) introduced in Sec. 4, is defined by the relation*

$$\operatorname{Ind} L^t = \chi(V') = -\operatorname{Ind} L, \tag{1}$$

where $\chi(V')$ *is an Euler characteristic of* V'.

REMARK. From the point of view of the theorem formulated above, Theorem 1, Sec. 4, describes the class of problems whose index assumes the values ± 1 in accordance with (1).

It stands to reason that the statement of the theorem formulated above is far from being obvious. We shall elucidate its meaning and consider the scheme of the proof.

Relation (1) is a consequence of the corresponding relation for $M = \overline{V}' \cup \overline{V}''$, but the transition requires the use of some auxiliary concepts and results. It turns out that the basis of the homology group for M in the dimensionality p consists of the cycles $\Gamma^i_{(p)}$, $i = 1, \ldots, B'_p$, which are cycles of V' nonhomologous to zero, and, in addition, of the cycles $\gamma^k_{(p)}$, $k = 1, \ldots, B''_p$ resulting from the "doubling" of V' (transition to M) from the so-called *relative* (not homologous to zero) cycles of V', generated by chains whose boundary belongs to that of V'.

Without considering strict definitions, we can point out that in Example 2c, 2.2, Ch. 1, this relative cycle is the element $x_1 \times \varepsilon$ (or $x_2 \times \varepsilon$) that generates a cycle on M (homeomorphic to the torus in this case), not homologous to the cycle $e_1 \times y_1 + e_2 \times y_1$.

The following variant of Hodge's theorem proves to be valid for the forms ω harmonic on M (satisfying the relations $d\omega = \delta\omega = 0$).

PROPOSITION 1. *For the given homology basis* $\{\Gamma^i_{(p)}\}$, $\{\gamma^k_{(p)}\}$ *on the manifold* M *of the structure having been described there exist even (odd) harmonic forms* $u^i_{(p)}$ ($v^k_{(p)}$, *respectively) such that*

$$\int_{\Gamma^i_{(p)}} u^j_{(p)} = \delta^j_i, \qquad \int_{\gamma^k_{(p)}} v^l_{(p)} = \delta^l_k, \qquad i, j = 1, \ldots, B'_p, \quad k, l = 1, \ldots, B''_p, \tag{2}$$

where δ_j^k *is the Kronecker delta. An even (odd) harmonic form, whose integrals over the cycles* $\Gamma^i_{(p)}$ *(or* $\gamma^k_{(p)}$*) are all equal to zero, is zero.*

This statement follows from the results given in [61, 65].

Since every solution of the homogeneous system (L), (L^t) is a harmonic form and, conversely, every nonzero harmonic form gives a nontrivial solution of the corresponding homogeneous system, Proposition 1 yields the following assertion.

PROPOSITION 2. *When we consider even forms, the necessary and sufficient condition for the solvability of the system* (L) *(or* (L^t)*) is the fulfillment of the conditions*

$$(f^i_{(p)}, u^i_{(p)}) = 0, \qquad i = 1, \dots, B'_p$$

for all even (odd) values of p. *When the additional conditions* $(\omega^i_{(p)}, u^i_{(p)}) = 0$, $i = 1, \dots, B'_p$, *are satisfied, where* p *assumes all odd (even) values, the solution* ω *is unique.*

Propositions 1, 2 yield the result of Theorem 1 that refers to even forms. Its validity after the transition to odd forms follows from the duality theorem ([30, p. 270]) for manifolds with boundary, namely, we have the following statement.

PROPOSITION 3. *The relations* $B'_p = B''_{n-p}$, $p = 0, 1, \dots, n$, *are valid for the manifold* V' *of dimension* n.

EXAMPLE 1. Let V' be a spherical layer $1 < r < 2$, $r^2 = x^2 + y^2 + z^2$. Let us define the forms

$$u_{(2)} = \frac{1}{4\pi} \frac{z\,dx \wedge dy - y\,dx \wedge dz + x\,dy \wedge dz}{r^3},$$

$$v_{(1)} = \frac{1}{2} \frac{x\,dx + y\,dy + z\,dz}{r}.$$

It can be directly verified that $u_{(2)}$ is continuous for an even and $v_{(1)}$ for an odd continuation to $M = \overline{V}' \cup \overline{V}''$. Being harmonic, they give forms for M which satisfy (2). Our standard problems have the index 2 in V'.

EXAMPLE 2. Let V' be an n-dimensional ball with N balls of a smaller radius lying entirely in V' removed. Then, for $\overline{V}' \cup \overline{V}'' = M$, we have

$$B'_0 = 1, \qquad B'_{n-1} = N,$$

and the other Betti numbers are equal to zero, i.e.,

$$\chi(V') = 1 + (-1)^{n-1}N.$$

Thus, for $N = 0, 1, 2, \ldots$, we get the values $\chi = 1, 0, -1, -2, \ldots$ for even n and $\chi = 1, 2, 3, \ldots$ for odd n. Also noting that $\chi(T^n) = 0$ for a torus (the example can be found in 3.2) and including the periodicity conditions when considering (L), (L^t) in a parallelepiped into the class of "standard" problems, we obtain the following statement.

PROPOSITION 4. *In a Euclidean space of an arbitrary dimension, for any integer m we can indicate a domain, a system of equations, and a class of boundary conditions such that the index of the corresponding problem will be equal to m.*

5.2. Orthogonal decompositions. Recall relation (8), Sec. 2:

$$\mathbb{H}^r = \mathfrak{R}_d \oplus \mathfrak{R}_\delta \oplus \mathfrak{N}_\Delta. \tag{3}$$

The constructions carried out in Sec. 4 allow us to obtain immediately a similar result for the space $\mathbb{H}^r(V')$, where V' is an arbitrary smooth manifold with boundary ($M = \overline{V}' \cup \overline{V}''$ is supposed to be compact as before).

Indeed, it is sufficient to assume that the corresponding constructions are carried out on M in subspaces of even or odd forms, i.e., the elements from $\mathcal{D}_d$, $\mathcal{D}_\delta$ (the domains of definition of the corresponding operators) are subject to the boundary conditions of form (6), (7), Sec. 4, that were used in the boundary value problems for (L), (L^t). From the considerations of 5.1 it follows that the subspace $\mathfrak{N}_\Delta$ will, in this case, be finite-dimensional and its dimension, according to whether the forms under consideration are even or odd, is defined by Betti numbers B'_p, B''_p.

It should also be pointed out that sometimes, on a manifold with boundary (in particular, in a domain of the Euclidean space) a different version of decomposition (3) (mentioned in Sec. 2) is used, in which the operators d, δ are initially defined on smooth finite forms and then their closure in $\mathbb{H}$ is taken. In this case (for a nondegenerate boundary) the subspace $\mathfrak{N}_\Delta$ turns out to be an infinite-dimensional subspace of $\mathbb{H}^r(V')$. Now the elements belonging to it are free from any boundary conditions and are harmonic at every interior point of V'.

5.3. The Cauchy integral. Let us return to the assumption that V is a domain of $\mathbb{R}^n$ with a sufficiently smooth boundary (we do not write V' since the double and the evenness of the forms do not participate in the discussion). It is now convenient to combine the systems (L), (L^t) and consider the collection of equations

$$K\omega \equiv (d+\delta)\omega = f, \tag{4}$$

in which ω, f are collections of forms of all degrees ($\omega = \omega_{(0)}, \ldots, \omega_{(n)}$; the same, similarly for f). We preserve the name of the form (*non-homogeneous* now) for this aggregate. As a result we have $K^t = K$ and, using relation (2), Sec. 1, we can write the Green formula for (4) in the form

$$(K\omega, x)_V = \int_{\partial V} [\omega \wedge *\chi - \chi \wedge *\omega] + (\omega, K\chi)_V. \tag{5}$$

This formula is basic for all subsequent considerations.

We say that ω is *holomorphic* if $K\omega = 0$. Let us introduce the holomorphic form e whose all components are unities. Assuming that ω is also holomorphic and setting $\chi = e$ in (5), we get

$$\int_{\partial U} [\omega \wedge *e - e \wedge *\omega] = 0 \tag{6}$$

for any subdomain $U_t \subset V$ (on the assumption of regularity of ∂U).

PROPOSITION 5 (the Cauchy theorem). *For the holomorphic form ω, an integral of form* (6) *of every one of its components $\omega_{i_1 \ldots i_p}(x)$ is zero.*

This statement implies that relation (6) is preserved if, instead of e, we take the form $e_{i_1 \ldots i_p}$, where the component with the indicated set of subscripts is unity and the other components are zeros. ■

We can give relation (6) a different form, which does not require any additional explanations. We introduce a *double form* [39]

$$e(x, y) = \sum_p \sum_{i_1 < \ldots < i_p} dx^{i_1} \wedge \ldots \wedge dx^{i_p} \wedge dy^{i_1} \wedge \ldots \wedge dy^{i_p},$$

supposing that dx^i, dy^j commute in the exterior multiplication and the operators d, δ, $*$, that act with respect to x (with respect to y), commute with dy^j (with dx^i). Then the relation

$$\chi(y) = \int_{\partial U} [\omega \wedge *_x e(x, y) - e(x, y) \wedge *_x \omega] = 0, \tag{7}$$

in which the integration is carried out with respect to the variables x means that all the coefficients of the form $\chi(y)$ vanish, and this is equivalent to Proposition 5.

The double form $e(x, y)$ is also convenient for writing the analog of the *Cauchy integral.* Let

$$\gamma(x, y) = e(x, y)r^{2-n}, \qquad r^2 = \sum_{k=1}^{n}(x^k - y^k)^2.$$

PROPOSITION 6. *If $\omega(x)$ is holomorphic in $V \supset \overline{U}$, then*

$$\omega(x) = c_n \int_{\partial U} [K_y\gamma(x, y) \wedge *_y\omega(y) - \omega(y) \wedge *_y K_y\gamma(x, y)] \tag{8}$$

(*integration with respect to y*) *for $x \in U$.*

In order to prove this statement, it is sufficient to set $\chi = K\gamma$ in (5) and note that $K^2\gamma = -\Delta\gamma = \delta_y(x)$ ("δ of Dirac"). ■

We can introduce by analogy an integral *of the type of "Cauchy integral".*

PROPOSITION 7. *If the form ω is continuous on ∂V, then the form*

$$\chi(x) = \int_{\partial V} [K_y\gamma(x, y) \wedge *_y\omega(y) - \omega(y) \wedge *_y K_y\gamma(x, y)] \tag{9}$$

(*integration with respect to y*) *is holomorphic in $\mathbb{R}^n \setminus \partial V$.* ■

In this case, we can obtain the jump formulas for χ and consider the representation $\chi(x)$ for $x \in \partial V$, which uses the **principal value** of integral (9) (cf. [3]).

The most nontrivial among the results of the type under consideration is the Morera theorem first formulated for $n = 3$ in [66] and then, for the general case, in [37] (the paper [37] also contains detailed proofs of propositions 5–7).

PROPOSITION 8 (Morera theorem). *Suppose that the form $\omega(x)$ is continuous in $V \subset \mathbb{R}^n$ and relation (7) is satisfied for it when integration is carried out along the boundary (regular enough) of any subdomain $U \subset V$. Then a representation of form (8) is valid for $\omega(x)$, i.e., it is holomorphic in V.* ■

6. Certain Equations of Mathematical Physics

6.0. Preliminary remarks. The Riemannian formalism, also known as a differential-geometric approach, which is sometimes extended to a complex case and is considerably more developed (using the concepts of curvature, connectedness, group-theoretic methods, etc.), is the working apparatus for a number of important lines of investigation of contemporary theoretic physics ([58], for instance). However, we shall restrict our discussion to purely classical examples that concern hydro- and electrodynamics.

In line with the general purposes set up in this book, the invariant interpretation of special problems is important, in particular, since it automatically leads to certain discrete analogs.

6.1. Euler equations. These equations describe the motion of the so-called ideal (nonviscous and incompressible) fluid. Their invariant treatment admits several different interpretations. We shall consider the interpretation suggested in [20].

In a two-dimensional stationary case, when there are no external forces, the classical form of Euler equations is

$$uD_x u + vD_y u + D_x p = 0, \qquad uD_x v + vD_y v + D_y p = 0,$$
$$D_x u + D_y v = 0, \tag{1}$$

where the pair of functions (u, v) defines the field of velocities and p is pressure. The last equation (that requires the equality of divergence to zero, i.e., incompressibility) is a consequence of the corresponding integral relation and immediately leads to the conclusion (cf. the beginning of 2.1) that we speak of the field of the covector $\mathfrak{u} = (u, v)$ and the function $p(x, y)$ is a scalar. The relation

$$d * \mathfrak{u} = 0 \tag{2}$$

(the solenoidality condition) is the invariant form of the last equation in (1), and this is equivalent to the relation

$$\delta\mathfrak{u} = 0. \tag{3}$$

It is natural to consider the second characteristic of the field $\mathfrak{u}$, namely, the 2-form $d\mathfrak{u}$. Following tradition, we introduce this characteristic setting

$$d\mathfrak{u} = *\omega, \tag{4}$$

where ω is the so-called *curl* of the field $\mathfrak{u}$ (in a two-dimensional case it must be a scalar and in a three-dimensional case, a covector).

We can regard Eq. (1) as equations defining a certain special relationship between $d\mathfrak{u}$ and $\delta\mathfrak{u}$, and we shall give the invariant notation of this relationship supposing that relations (2), (3), (4) are written for the covector $\mathfrak{u}$ in an n-dimensional case.

We call the form $\chi \in \Lambda^{n-2}$ a *multiplier* of the field $\mathfrak{u}$ if

$$d * (\mathfrak{u} \wedge \chi) = 0,$$

i.e., if a relation of the form (2) is again satisfied for $\mathfrak{u} \wedge \chi$.

REMARK. The field $\mathfrak{u}$ is *Eulerian* if its curl is a multiplier at the same time, i.e., if

$$d * (\mathfrak{u} \wedge \omega) = \pm d * (\mathfrak{u} \wedge * \, d\mathfrak{u}) = 0 \tag{5}$$

(we have used the fact that $\pm\omega = * \, d\mathfrak{u}$).

In a simply connected domain, (5) implies the existence of a scalar γ such that

$$* \, (\mathfrak{u} \wedge * \, d\mathfrak{u}) = -d\gamma \tag{6}$$

(the minus sign is used by convention). For $n = 2$, in the coordinate notation, (6) gives

$$\begin{aligned} -v(D_x v - D_y u) + D_x \gamma = 0, \\ u(D_x v - D_y u) + D_y \gamma = 0, \end{aligned} \tag{7}$$

and this corresponds to the so-called Gromeko–Lamb form [28] of Euler equations. The transition from (1) to (7) corresponds to the substitution $\gamma = p + \frac{1}{2}(u^2 + v^2)$. The scalar γ is sometimes called the **dynamic pressure**. A similar transition from (6) to the corresponding coordinate form of Euler equations can also be performed for $n = 3$.

Relation (5), which is a consequence of Euler equations often used in hydrodynamics, is one of the forms of the "law of preservation of curl". Let us consider the invariant notation of some other classical relations. Applying the operator $*$ to (6) and using the (external) multiplication by $\mathfrak{u}$, we obtain

$$\mathfrak{u} \wedge * \, d\gamma = 0,$$

and this corresponds to the **Bernoulli theorem**. We can obtain the simplest nontrivial solution of Eq. (5) setting $d\mathfrak{u} = 0$. Flows of this kind are said to be *potential.*

The solenoidality (the validity of relation (2)) is equivalent, at least locally, to the existence of the form $\psi \in \Lambda^2$ such that $\mathfrak{u} = \delta\psi$ (we have used (3) instead of (2)). In a two-dimensional case, ψ ($*\,\psi$, to be more precise) is a classical stream function; for $n = 3$ it is a "vector". Noting that we can always suppose that $d\psi = 0$ (for $n = 2$ it is automatically satisfied) and making use of (4), we obtain

$$-\Delta\psi = (d\delta + \delta d)\psi = *\,\omega.$$

Also note that for $n = 2$ (5) implies that $d\psi \wedge d\omega = 0$, and this gives the relationship between the functions ω and ψ (the equality of the corresponding Jacobian to zero).

Some remarks are due concerning a nonstationary case, in which the standard form of the equations is

$$\frac{\partial \mathfrak{u}}{\partial t} + *\,(\mathfrak{u} \wedge *\, d\mathfrak{u}) + d\gamma = 0. \tag{8}$$

From this we can obtain one of the classical conservation laws. We apply the operator $*$ to (8) and use the multiplication on the left (externally) by $\mathfrak{u}$:

$$\mathfrak{u} \wedge * \frac{\partial \mathfrak{u}}{\partial t} + \mathfrak{u} \wedge *\, d\gamma = 0. \tag{9}$$

However, $\mathfrak{u} \wedge *d\gamma = d\gamma \wedge *\mathfrak{u}$, and, at the same time,

$$\int\limits_V d\gamma \wedge *\,\mathfrak{u} = \int\limits_V d(\gamma \wedge *\,\mathfrak{u}) - \int\limits_V \gamma \wedge d *\mathfrak{u}. \tag{10}$$

The last term in (10) is zero because of $d{*}\mathfrak{u} = 0$ and the last but one term (after the transition to the integral over ∂V) vanishes provided that $\gamma|_{\partial V} = 0$ or $*\,\mathfrak{u}|_{\partial V} = 0$. Thus, integrating (9) over the domain V and using the remarks that have been made, we obtain

$$\int\limits_V \mathfrak{u} \wedge \frac{\partial}{\partial t} *\,\mathfrak{u} = \frac{1}{2}\frac{\partial}{\partial t}\int\limits_V \mathfrak{u} \wedge *\mathfrak{u} = \frac{1}{2}\frac{\partial}{\partial t}(\mathfrak{u}, \mathfrak{u})_V = 0.$$

It remains to integrate this relation in the limits from 0 to T.

On the other hand, applying the operator d to (8), and then the operator $*$, we get the so-called Helmholtz equation for the curl:

$$\frac{\partial}{\partial t}\omega + * d * (\mathfrak{u} \wedge \omega) = 0$$

(we do not replace $* d *$ by δ in order to obviate the indeterminacy in the sign). It should be emphasized that the formalism we have introduced automatically gives the notation of the analogs of classical equations in the Euclidean space (or on a manifold) of an arbitrary dimension.

At the same time, it should be pointed out that the presented discussion does not exhaust the possibilities of the differential-geometric approach to Eqs. (1). Thus, Eqs. (8) can be written in the form

$$\frac{d\omega}{dt} = B(\omega, \omega),$$

where ω is an element of a certain Lie algebra $\mathfrak{a}$ and $B: \mathfrak{a} \times \mathfrak{a} \to \mathfrak{a}$ is a certain bilinear operation [59]. If $\mathfrak{a}$ is the Lie algebra of the group $SO(3)$, then the equation written out is the classical system of Euler equations which describe the motion of a rigid body with a fixed point, in which ω is the angular velocity vector in the coordinate system rigidly connected with the body.

If we use the fact that the terms of (1), which contain the products of u, v by the derivatives, can be written as the Lie derivative [47] of $\mathfrak{u}$ with respect to the vector $J\mathfrak{u}$, $J : \Lambda^1 \to T^1$, then the equations (7) can be written as

$$(J\mathfrak{u}) \lrcorner d\mathfrak{u} + d\gamma = 0,$$

where $\lrcorner$ is the operation of the so-called interval multiplication of a vector by a covector [47]. This form of notation was used in [16].

6.2. Certain special flows. Here we shall touch upon the following two problems: first, the usefulness of the invariant form of notation of equations in the transition to coordinate systems different from the Cartesian system (i.e., to a metric different from the Euclidean one); second, some methods of studying the structural properties of Euler flows (i.e., flows described by the solutions of Eqs. (1)). The first theme corresponds to the content of 1.2 and the second, to that of Sec. 4, Ch. 3.

EXAMPLE 1. Let us consider Eqs. (1) on a two-dimensional sphere with radius r. Using the ordinary notation of spherical coordinates,

$$x = r\cos\varphi, \qquad y = r\sin\varphi\cos\vartheta, \qquad z = r\sin\varphi\sin\vartheta,$$

we can consider the sphere to be parametrized by the variables

$$0 \le \varphi \le \pi, \qquad 0 \le \vartheta < 2\pi.$$

Representing $\mathfrak{u}$ as

$$\mathfrak{u} = u(\varphi,\vartheta)d\varphi + v(\varphi,\vartheta)d\vartheta$$

and noting that the metric tensor has components

$$g_{11} = r^2, \qquad g_{22} = r^2\sin^2\varphi, \qquad g_{12} = 0,$$

we can write (see Sec. 1)

$$*\mathfrak{u} = \frac{1}{\sin\varphi} v\,d\varphi - \sin\varphi\, u\, d\vartheta$$

(as would be expected, in this case $\mathfrak{u} \wedge *\mathfrak{u} = (u^2+v^2)r^2\sin\varphi)$. Using (2) and introducing the stream function $*\mathfrak{u} = d\psi$ (deviating somewhat from the notations used in 6.1, we do not use the operator δ), we obtain

$$\psi_\varphi = \frac{v}{\sin\varphi}, \qquad \psi_\vartheta = (-\sin\varphi)u.$$

The relationship between the curl and the stream function, $-\Delta\psi = \omega$, assumes the form

$$D_\varphi(\psi_\varphi\sin\varphi) + D_\vartheta\Big(\frac{\psi_\vartheta}{\sin\varphi}\Big) = r^2\omega\sin\varphi. \tag{11}$$

Relation (5) immediately leads to the relation

$$\psi_\varphi\omega_\vartheta - \psi_\vartheta\omega_\varphi = 0,$$

which has an invariant character and the same form as in a two-dimensional case.

Equation (11) on the surface of a sphere is solvable only under the condition of orthogonality of ω to the constants, i.e., $\int *\omega = 0$. Correspondingly, the simplest Euler flow is the rotation of a "fluid" as a whole, i.e., $\psi(\varphi, \vartheta) = 0$, $u = 0$, $v = -\sin\varphi$, $\omega = -r^{-2}\cos\varphi$, $\varphi \neq 0, \pi$.

EXAMPLE 2. Let us consider the axially symmetric flows ([34, Ch. 15]). Suppose that (x, r, ϑ) are cylindrical coordinates ($y = r\cos\vartheta$, $z = r\sin\vartheta$) and the x-axis is the axis of symmetry of the flow:

$$\mathfrak{u} = u(x,r)dx + v(x,r)dr + w(x,r)d\vartheta.$$

It is customary to understand an axially symmetric flow as a plane axially symmetric flow, i.e., set $w \equiv 0$. Again, in accordance with Sec. 1,

$$*\,\mathfrak{u} = \frac{1}{r} w\, dx \wedge dr - rv\, dx \wedge d\vartheta + ru\, dr \wedge d\vartheta. \tag{12}$$

Relation (2) yields $ru_x + rv_r + v = 0$, i.e., the classical form of the notation of the equality of the divergence to zero in the cylindrical coordinates [34]. At the same time, (2) implies the existence (at least locally) of the covector $\psi = \psi_1\, dx + \psi_2\, dy + \psi_3\, dz$, for which $d\psi = *\,\mathfrak{u}$. Taking into account that $\psi_{1,\vartheta} = \psi_{2,\vartheta} = 0$, we find from (12) that

$$\psi_{3,r} = -rv, \qquad \psi_{3,x} = ru, \qquad \psi_{2,x} - \psi_{1,r} = 0.$$

The last relation is of no particular interest and the first two relations imply the existence of the so-called Stokes function $S(x,r) = \psi_3(x,r)$ such that

$$\mathfrak{u} = \frac{1}{r} S_r\, dx - \frac{1}{r} S_r\, dr.$$

Introducing the curl in accordance with (4), we see that (4) turns out to be equivalent to the scalar relation

$$-S_{xx} - S_{rr} + \frac{1}{r} S_r = \omega.$$

Condition (5) assumes the form

$$S_x \omega_r - S_r \omega_x - \frac{2}{r} S_x \omega = 0.$$

In addition to these examples, we should note that the ordinary procedures of seeking the form of $-\Delta$ in spherical coordinates are also special cases of the considerations we have carried out (the so-called Lamé parameters are nothing other than the square roots of the components g_{11}, g_{22}, g_{33} of a metric tensor).

Let us pass to the second part, i.e., to the remarks concerning the study of Eqs. (1). One of the main causes of the difficulties encountered here is the absence of the sufficiently general theorems on the existence and uniqueness of solutions (cf. [49]). The following statement illustrates the phenomena causing this situation.

PROPOSITION 1. *Suppose that $r_{\mathcal{P}}(x,y)$ is a Euclidean distance to the fixed point $\mathcal{P} \in \mathbb{R}^2$ and f is an arbitrary smooth function defined on $[0,\infty)$, $f(0) \equiv 0$. Then the flow, defined by the function of current $\psi(x,y) = f(r_{\mathcal{P}})$ is Eulerian.*

PROOF. It is sufficient to consider the case where $\mathcal{P} = 0$. We have

$$u = f_r'\frac{y}{r}, \qquad v = -f_r'\frac{x}{r}, \qquad \omega = -\Delta\psi = -f_{rr}'' - \frac{f_r'}{r},$$

and can write the representation

$$v\omega = \varphi(r)\frac{x}{r}, \qquad -u\omega = \varphi(r)\frac{y}{r},$$

that makes it possible to assert the existence of the function $\gamma(x,y)$ such that Eqs. (7) are satisfied. ■

COROLLARY. *In the arbitrary bounded domain V there exists a continuum of various Euler flows that satisfy the conditions $u = v = p = 0$ on the boundary of V.*

Indeed, in order to obtain the corresponding Euler flow, it is sufficient to suppose the function f from Proposition 1 to be finite in V.

The most frequently used method of excluding from the considerations the flows of the type having been described is the use of the requirement of potentiality or the requirement of the absence of singular points of the field $\mathbf{u}$ (points, where $u^2 + v^2 = 0$). For non-potential flows the field $\mathbf{u}$ uniquely defined in V can be defined by means of specifying ω and defining ψ from the equation $-\Delta\psi = \omega$, to which the boundary conditions for ψ on the boundary of V are added. The problem of finding the corresponding flows becomes linear, but not for any choice of ω is the flow that we have found, Eulerian. In a plane case, a flow is Eulerian if one of the relations,

$$\omega = \mu(\psi) \quad \text{or} \quad \psi = \nu(\omega), \tag{13}$$

following from (5), is valid. However, the requirement of the validity of both relations is not necessary.

PROPOSITION 2. *The Euler flow corresponding to the functions*

$$\psi = x - x^4, \qquad \omega = 12x^2,$$

does not admit the first relation in the neighborhood of the line $x = 4^{-1/3}$ *and the second relation in the neighborhood of the line* $x = 0$.

The verification is elementary. ■

The vanishing, equivalent to (13), of the corresponding Jacobian allows us simultaneously to verify the following statement.

PROPOSITION 3. *There does not exist an Eulerian flow with the curl* $\omega = xy$ *in the domain* $1 > y > x > 0$. ■

Similar constructions are possible in an axially symmetric case as well. The problem of the admissibility of a curl is essentially more difficult when we pass from local considerations to boundary value problems.

EXAMPLE. Suppose that $V = \{|x| \leq \alpha, |y| \leq \pi/2\}$ and $\omega = \beta \cos y$, $\beta = \text{const}$. Let us find the solutions of the equation

$$-\Delta\psi = \omega = \beta \cos y, \tag{14}$$

adding to it the boundary conditions

$$\psi(-\alpha, y) = -\cos y, \qquad \psi(\alpha, y) = \psi(x, \pm\pi/2) = 0.$$

The corresponding solution is unique and is given by the function

$$\psi(x, y) = \xi(x)\cos y, \qquad -\xi'' + \xi = \beta, \qquad \xi(-\alpha) = -1, \quad \xi(\alpha) = 0.$$

The corresponding field $u = \psi_y$, $v = -\psi_x$ is not Eulerian for any value $\beta \neq 0$. At the same time, if we remove the boundary conditions, for any β there will be functions of current ψ which define Euler flows. For instance, $\psi = \beta \cos y$, $\beta = \text{const}$.

This range of problems has not yet been thoroughly investigated.

6.3. The Navier–Stokes equations and linearization. In a plane stationary case, the Navier–Stokes equations can be obtained from (1) by adding the terms $-\mu\Delta u$, $-\mu\Delta v$, respectively, to the left-hand side of the first and second equations. The constant $\mu \geq 0$ characterizes the viscosity. Since the invariant notation of the operator $-\Delta$ is known, every invariant form of Eqs. (1) gives the invariant form of the Navier–Stokes equations. We shall not study these equations more thoroughly since there is extensive literature devoted to this theme. I shall only make some remarks which establish the relationship between the simplified modifications (linearizations) and our discussion.

The simplest variant results when the quasilinear ("convective") terms are rejected. This gives, in our notations,

$$-\mu\Delta\mathfrak{u} + dp = 0, \qquad \delta\mathfrak{u} = 0. \tag{15}$$

When we study system (15), the orthogonal decomposition of the space $\overline{\mathbb{H}}(V)$ of the type considered in 2.2, 5.2 often turn out to be convenient. Thus, if we fix a certain domain V, and $\overline{\mathbb{H}}(V) = \mathfrak{R}_d \oplus \mathfrak{N}_\delta$, then the corresponding problem for (15) reduces to the investigation of the equation $[-\mu\Delta u]_{\mathfrak{N}_\delta} = 0$ for the projection. The boundary conditions can be included into the definition of $\mathcal{D}(d)$ and in the definition of $\mathcal{D}(\delta)$.

Another simplification results from the replacement of the convective terms $\mathfrak{u} \wedge * d\mathfrak{u}$ by $A \wedge * d\mathfrak{u}$ or $\mathfrak{u} \wedge * dA$, where the covector A is supposed to be known. After this replacement the Navier–Stokes equations become linear again.

Such a substitution is often also performed in Euler equations we have considered. The resulting first-order systems turn out to be symmetric, and symmetric positive in a nonstationary case. Thus, as a result of the replacement of the quasilinear terms by $*(A \wedge * d\mathfrak{u})$, we have, in matrix notation (a two-dimensional nonstationary case),

$$L \equiv \begin{pmatrix} 1 & 0 & 0 \\ 0 & 1 & 0 \\ 0 & 0 & \sigma \end{pmatrix} D_t + \begin{pmatrix} a_1 & 0 & 1 \\ 0 & a_2 & 0 \\ 1 & 0 & 0 \end{pmatrix} D_x + \begin{pmatrix} -a_1 & 0 & 0 \\ 0 & -a_2 & -1 \\ 0 & 1 & 0 \end{pmatrix} D_y,$$

where L is applied to the column (u, v, p). In the case that we considered in 6.1 we had $\sigma = 0$. The transition to $\sigma = 1$ corresponds to the transition to a symmetric hyperbolic system that describes the flow of compressible fluid.

In a three-dimensional nonstationary case, choosing the unit vector $(0,0,1)$ as $*dA$ in the group of terms $*(\mathfrak{u} \wedge * dA)$, we get the familiar linear system which describes the motion of rotating fluid [44].

The methods described above (cf. [12, 14]) are applicable for obtaining the theorems on the existence and uniqueness of solutions of boundary value problems for first-order systems of the type having been described. Sometimes these methods are used in conjunction with the methods of orthogonal decompositions.

6.4. Maxwell equations. By the way of another example we shall consider the invariant notation of the Maxwell equations. An electromagnetic field can be described by the definition of the "vector potential" A (cf. [4]), which is a covector in a four-dimensional space with a Lorentz metric, defined by a metric tensor in which only the components $g_{11} = g_{22} = g_{33} = -g_{44} = 1$ are nonzero. The potential A must satisfy the equation

$$\delta\, dA = f, \tag{16}$$

in which the covector f describes the charges and currents and is subjected to the compatibility condition $\delta f = 0$. If we introduce the 2-form $\mathcal{F}$ (in physical treaties called a skew-symmetric tensor of rank 2) defined by the relation $\mathcal{F} = dA$, then (16) can be replaced by the pair of equations

$$d\mathcal{F} = 0, \qquad \delta\mathcal{F} = f. \tag{17}$$

This is the so-called *four-dimensional form of the Maxwell equations.*

If we now represent the space-time domain V, in which we consider the equations, in the form $V = Q \times T$, where Q is the space domain of the variables $x = (x^1, x^2, x^3)$ and T is the time interval $x^4 \equiv t \in T$, then we can apply to (17) the splitting mechanism described in 3.4, i.e., represent $\mathcal{F}$ as

$$\mathcal{F} = u_{\mathrm{II}} \otimes v_{(0)} + u_{\mathrm{I}} \otimes v_{(1)}.$$

This is a version of relation (12) from Sec. 3. Correspondingly (cf. (13), Sec. 3), we obtain

$$d\mathcal{F} = du_{\mathrm{II}} \otimes v_{(0)} + u_{\mathrm{II}} \otimes dv_{(0)} + du_{\mathrm{I}} \otimes v_{(1)},$$

$$\delta\mathcal{F} = \delta u_{\mathrm{II}} \otimes v_{(0)} + \delta u_{\mathrm{II}} \otimes v_{(1)} + u_{\mathrm{I}} \otimes \delta v_{(1)}.$$

We have changed the sign in the last term of the second row, having taken into account that $g_{44} = -1$. Using the corresponding splitting for f,

$$f = h_{(1)} \otimes g_{(0)} + h_{(0)} \otimes g_{(1)} = 4\pi j - 4\pi\rho,$$

where j is a covector (over Q) and ρ is a scalar, we can write (17) as

$$\begin{aligned} &u_{\rm II} \otimes dv_{(0)} + du_{\rm I} \otimes v_{(1)} = 0, \qquad du_{\rm II} \otimes v_{(0)} = 0, \\ &u_{\rm I} \otimes \delta v_{(1)} + \delta u_{\rm II} \otimes v_{(0)} = 4\pi j, \quad \delta u_{\rm I} \otimes v_{(1)} = -4\pi\rho, \end{aligned} \tag{18}$$

where j defines the currents and ρ the charges.

Recall that when using this formalism, we must bear in mind the remark to (12) in Sec. 3. It should also be pointed out that since $\mathcal{F}$ is a 2-form, as distinct from the general considerations of 3.4, $u_{\rm II}$ consists only of one 2-form $u_{\rm II}(x)$ and $u_{\rm I}$ consists of one 1-form $u_{\rm I}(x)$.

In order to pass from (18) to the ordinary vector notations, we make a substitution $u_{\rm II} = *w_{\rm I}$ and set

$$w \otimes v_{(0)} = H, \qquad u_{\rm I} \otimes v_{(1)} = E.$$

We obtain

$$\begin{aligned} &D_t H = -* dE, \quad D_t E = * dH - 4\pi j, \\ &d * H = 0, \qquad \delta E = -4\pi\rho, \end{aligned} \tag{19}$$

where the operators d, δ, $*$ act over Q and the action of d, δ over T (i.e., $dv_{(0)}$, $\delta v_{(1)}$) is replaced by D_t, $-D_t$, respectively. If we now take into account that

$$* d = \text{rot}\,, \qquad d* = -\delta = \text{div}\,,$$

then it becomes clear that (19) is a standard notation of the Maxwell equations.

System (19) is overdetermined and not sufficiently studied in the general case. The standard technique of simplification is the differentiation of the first equation of (19) with respect to t and exclusion of $D_t E$ with the use of the second equation. Supposing that $j = 0$, we obtain

$$D_t^2 H = - * d * dH \equiv -\delta dH = -\Delta H$$

(since $d\delta H = 0$), i.e., a classical wave equation.

Now if we reject the last two equations in (19), we get a symmetric hyperbolic system which can be investigated by means of the corresponding techniques mentioned above.

It would possibly be of interest to study (17) with the use of the nonstandard splitting $V = Q_2 \times \mathcal{L}_2$ into 2 two-dimensional domains with a Euclidean metric in the first and a Lorentz metric in the second. In this case,

$$\mathcal{F} = u_{(0)} \otimes v_{(2)} + u_{(1)} \otimes v_{(1)} + u_{(2)} \otimes v_{(0)},$$

where the subscript is the rank of the corresponding form.

It should be added that the described approach to Maxwell equations admits an important generalization, which leads immediately to the so-called Yang–Mills equations of the quantum field theory. In this case, the range of 2-form A belongs to the Lie algebra $SU(2)$, and A defines a connection on the corresponding fiber space [68].

Chapter 3

The Model of Euclidean Space and Difference Operators

0. Introductory Remarks

This chapter is the central one, and not only in the system of numbering. It is the attempt to give a coherent exposition of the constructions and results it contains that served as a stimulus for writing this monograph.

In Sec. 1, I construct an infinite complex whose special structure allows us to regard it as a combinatorial analog of the Euclidean space. It is implied that this structure makes it possible to define the analog of the metric operation $*$ and achieve the far-reaching parallelism between discrete and continual constructions. This parallelism is studied in Sec. 2, which repeats, in a certain sense, the content of Secs. 1–4 of Ch. 2. The third section is devoted to the pedantic study of a two-dimensional case, including the limiting process.

The detailed analysis of the difference analog of the method of orthogonal decompositions (which will be given schematically in Sec. 2) should be specially emphasized since it is known that the projection of the vectors $\mathfrak{u}$, which satisfy the additional condition $\operatorname{div} \mathfrak{u} = 0$, causes much difficulty in computations.

The fourth section contains the discretization of equations and of certain relations following from them, which were presented in the hydrodynamic part of Sec. 6, Ch. 2. The fifth section is devoted to a brief description of an alternative approach to the combinatorial modeling of the Euclidean structures, an approach which has some advantages as well as drawbacks.

Since there are a large number of problems, not sufficiently studied, which are connected with the construction of models, they are briefly described in the concluding section.

Recall finally that it was mentioned in the Preface that the use of ideas related to the material given in this chapter reflects the trends of the contemporary computational practice (see [41], for instance).

1. The Combinatorial Model of the Euclidean Space

1.0. Preliminary remarks. This section is devoted to formal constructions on which the content of Secs. 2–4 is based. Sec. 2 is the principal one, which deals with the tensor product of "combinatorial lines" that gives the model of the Euclidean space. The content of 1.3 is separated from the whole text and is not used anywhere in the sequel. It is devoted to the "topological" limiting process which corresponds to the refinement of the combinatorial structure that is not based on metric characteristics.

1.1. The model of the real line. Here we shall repeat the gist of the first part of Sec. 1, Ch. 0, proceeding, however, from the definitions given in Sec. 2, Ch. 1.

We introduce the sets $\{x_k\}$, $\{e_k\}$, $k \in \mathbb{Z}$, which we shall regard as the generators of the free Abelian groups of zero-dimensional and one-dimensional chains of the one-dimensional complex $\mathfrak{C} = \mathfrak{C}^0 \oplus \mathfrak{C}^1$ with the boundary homomorphism ∂,

$$\partial : \mathfrak{C}^1 \to \mathfrak{C}^0; \quad e_k \mapsto x_{\tau k} - x_k, \qquad \partial : \mathfrak{C}^0 \to 0.$$

We use here the shift operator $\tau k = k + 1$, $\tau^{-1} = \sigma$, introduced in Ch. 0. The free Abelian group is understood as the direct sum of infinite cyclic groups generated by $\{x_k\}$, $\{e_k\}$. The geometric interpretation of $\mathfrak{C}$ was considered in detail in 1.1, Ch. 0. In contrast to Ch. 0, the chains

$$a = \sum a^k x_k, \qquad b = \sum b^k e_k$$

are now chains with **integral** coefficients, and this is more natural ($\mathfrak{C}^0$, $\mathfrak{C}^1$ are Abelian groups and not vector spaces). We call the complex $\mathfrak{C}$ a *model of the real line.*

As we saw in Ch. 0, the investigation and use of this model are closely connected with the consideration of the dual formation, namely, the "functional space". The conjugate complex $K = K^0 \oplus K^1$ with the base elements $\{x^k\}$, $\{e^k\}$, considered in Ch. 0, is, in the terms of 2.3, Ch. 1, a one-dimensional complex of cochains with real coefficients. For the homomorphisms

$$\langle x_i, x^k \rangle = \langle e_i, e^k \rangle = \delta_i^k$$

we retain the term of the *pairing operation.* The cochains

$$\alpha = \sum \alpha_k x^k, \qquad \beta = \sum \beta_k e^k$$

will be called forms as before. We introduce the coboundary operator d by the relation

$$\langle \partial b, \alpha \rangle = \langle b, d\alpha \rangle,$$

setting $d: K^1 \to 0$. As in Ch. 0, we introduce in K the multiplication $\smile$ (Whitney multiplication [30]) setting

$$x^k \smile x^k = x^k, \qquad x^k \smile e^k = e^k \smile x^{\tau k} = e^k,$$

considering the product to be equal to zero in all other cases, and linearly extending it to arbitrary cochains.

PROPOSITION 1. *The relation*

$$d(\varphi \smile \psi) = d\varphi \smile \psi + \varphi \smile d\psi$$

is valid for the cochains $\varphi, \psi \in K$.

We have carried out the verification in Ch. 0. ■

We introduce the operation $*: \mathfrak{C}^0 \to \mathfrak{C}^1$, $\mathfrak{C}^1 \to \mathfrak{C}^0$, setting

$$* x^k = e^k, \qquad * e^k = x^{\tau k}.$$

REMARK. It should be pointed out that here a demarcation line passes between the standard constructions of algebraic topology (that we have used in their primitive form) and a special formalism which is convenient for simulating the analysis problems. The Whitney multiplication belongs to algebraic topology and the $*$ operation; its use in the definition of an inner product and the operator δ belongs to a special formalism which models the Riemannian structure.

Just as in Ch. 0, if now $V = \sum_1^N e_k$, then

$$(\varphi, \psi)_V = \langle V, \varphi \smile * \psi \rangle$$

is an inner product over V that defines the Hilbert spaces $\mathbb{H}^0$, $\mathbb{H}^1$. The relation

$$(d\alpha, \beta)_V = \langle \partial V, \alpha \smile * \beta \rangle + (\alpha, \delta\beta)_V$$

(relation (16) in Sec. 1, Ch. 0) defines the operator $\delta : \ K^1 \to K^0$, $\delta\beta = - *^{-1} d * \beta$, which is a metric conjugate of d.

If we return to Sec. 1, Ch. 2, we shall see that we have simulated a number of important relations typical of the Riemannian structure.

1.2. A model of the n-dimensional Euclidean space. We consider the tensor degree $\mathfrak{C}(n) = \otimes_1^n \mathfrak{C}$ of a one-dimensional complex, i.e., of the model of the real line described above, to be such a model. The definition of the tensor product of complexes was given in 2.4, Ch. 1. It was also pointed out that the complex of cochains $K(n)$, which is a conjugate of $\mathfrak{C}(n)$, with the operator d defined in it has a similar structure $K(n) = \otimes_1^n K$. Only the definition and the elucidation of the properties of the operations $\smile$ and $*$ in $K(n)$ require an additional consideration. In order to carry out the requisite constructions, we shall use induction on the dimension n and the results, given in 1.1, referring to a one-dimensional case.

In this case, it is convenient to write the base elements of the complex $K(n+1)$ in the form $s^{(p)} \otimes s^{\varkappa}$, where $s^{(p)}$ is a base element of $K(n)$ and $s^{\varkappa}$ is either $x^{\varkappa}$ or $e^{\varkappa}$, $\varkappa \in \mathbb{Z}$.

REMARK. It should be emphasized that whereas the superscript $\varkappa$ in the notation of $x^{\varkappa}$, $e^{\varkappa}$ is simply an integer, the structure of the superscript (p) is complicated enough. For example, for $n = 3$, the "point" $(\varkappa_1, \varkappa_2, \varkappa_3)$ (corresponding to the zero-dimensional base element $x^{\varkappa_1} \otimes x^{\varkappa_2} \otimes x^{\varkappa_3}$, $\varkappa \in \mathbb{Z}$) is associated, besides the zero-dimensional element, with 3 one-dimensional elements

$$x^{\varkappa_1} \otimes x^{\varkappa_2} \otimes e^{\varkappa_3}, \qquad x^{\varkappa_1} \otimes e^{\varkappa_2} \otimes x^{\varkappa_3}, \qquad e^{\varkappa_1} \otimes x^{\varkappa_2} \otimes x^{\varkappa_3},$$

3 two-dimensional elements (in which the factor $e^{\varkappa_i}$ appears twice), and 1 three-dimensional element. It is easy to realize that this corresponds to the number of components of the forms $\omega_{(r)}(x)$, $r =$

0, 1, 2, 3, at the point with the coordinates (x_1, x_2, x_3) whose part is played by the triple $(\varkappa_1, \varkappa_2, \varkappa_3)$. It is supposed that the whole requisite information (the coordinate of the "point", the dimension, and the number of "components") is contained in the symbol (p). This analysis obviously also refers to the structure of the base elements $s_{(p)}$ of the complex $\mathfrak{C}(n)$.

Supposing that the $\smile$-multiplication in $K(n)$ has been defined, we introduce it for the base elements of $K(n+1)$ according to the law

$$(s^{(p)} \otimes s^{\varkappa}) \smile (s^{(q)} \otimes s^{\mu}) = Q(\varkappa, q)(s^{(p)} \smile s^{(q)}) \otimes (s^{\varkappa} \smile s^{\mu}), \quad (1)$$

where the signum function $Q(\varkappa, q)$ is equal to -1 if the dimension of both elements $s^{\varkappa}$, $s^{(q)}$ is odd and to $+1$ otherwise. Since the multiplication is defined for $n = 1$, relation (1) extends the definition to the case of an arbitrary n. The extension of the multiplication to arbitrary forms is linear.

Let us establish the relations between the multiplication introduced and the operator d.

PROPOSITION 2. *The relation*

$$d(s \smile t) = ds \smile t + Q(s)s \smile dt, \quad (2)$$

where $Q(s) = (-1)^{\dim s}$, *is valid for* $s, t \in K(n)$.

In order to prove this statement, we shall use induction on n. For $n = 1$ relation (2) coincides with Proposition 1 (the absence of the signum function in the latter is a consequence of a trivial situation). Considering (2) to be valid for $K(n)$, let us verify it for $K(n+1)$. It is sufficient to carry out the verification for the base elements. We set

$$s = s^{(p)} \otimes s^{\varkappa}, \qquad t = s^{(r)} \otimes s^{\mu}.$$

In order to simplify the use of the sign rule when we use (1) and (12), Sec. 2, Ch. 1 (the definition of d in a tensor product), it is convenient to consider two cases separately, namely, $s^{\varkappa} = x^{\varkappa}$, and $s^{\varkappa} = e^{\varkappa}$. We shall consider in detail the second case since the first case is somewhat simpler.

We compare two chains of relations. On one hand,

$$d[(s^{(p)} \otimes e^{\varkappa}) \smile (s^{(r)} \otimes s^{\mu})] = Q(r)d[(s^{(p)} \smile s^{(r)}) \otimes (e^{\varkappa} \smile s^{\mu})] =$$

$$Q(r)d(s^{(p)} \smile s^{(r)}) \otimes (e^{\varkappa} \smile s^{\mu}) + 0 =$$

$$[Q(r)ds^{(p)} \smile s^{(r)} + Q(r)Q(p)s^{(p)} \smile ds^{(r)}] \otimes (e^{\varkappa} \smile s^{\mu}). \qquad (3)$$

Here we used twice the one-dimensionality of $e^{\varkappa}$, namely, replaced $Q(r, \varkappa)$ by $Q(r)$ and noted that $d(e^{\varkappa} \smile s^{\varkappa})$ always vanishes.

On the other hand, using the one-dimensionality of $e^{\varkappa}$ again, we have

$$d(s^{(p)} \otimes e^{\varkappa}) \smile (s^{(r)} \otimes s^{\mu}) - Q(p)(s^{(p)} \otimes e^{\varkappa}) \smile d(s^{(r)} \otimes s^{\mu}) =$$
$$Q(r)(ds^{(p)} \smile s^{(r)}) \otimes (e^{\varkappa} \smile s^{\mu}) + 0 -$$
$$Q(p)(s^{(p)} \otimes e^{\varkappa}) \smile (ds^{(r)} \otimes s^{\mu} - Q(r)s^{(r)} \otimes ds^{\mu}). \qquad (4)$$

The first term on the right-hand side of (4) coincides with the first term in the last row of (3) resulting from the removal of the parentheses. In the last term on the right-hand side of (4) only the product $-Q(p)(s^{(p)} \otimes e^{\varkappa}) \smile (ds^{(r)} \otimes s^{\mu})$ does not vanish. If $ds^{(r)} \neq 0$, it reduces to the form

$$Q(r)Q(p)(s^{(p)} \smile ds^{(r)}) \otimes (e^{\varkappa} \smile s^{\mu}),$$

i.e., coincides with the corresponding term in the last row of (3). We obviously have this coincidence in the case of $ds^{(r)} = 0$ as well. We have completed the verification of (2) for $s^{\varkappa} = e^{\varkappa}$. The verification for $s^{\varkappa} = x^{\varkappa}$ is similar and even somewhat simpler. ∎

PROPOSITION 3. *Suppose that $s^{(p)}$ is a base element of $K(n)$ defined at a certain "point" $\varkappa = (\varkappa_1, \dots, \varkappa_n)$. Then there exists a single element*

$$* s^{(p)}, \qquad \dim s^{(p)} + \dim *s^{(p)} = n$$

such that $s^{(p)} \smile * s^{(p)} = V^{\varkappa} = e^{\varkappa_1} \otimes \ldots \otimes e^{\varkappa_n}$. ∎

This statement defines the operation $*$. Its validity is obvious enough although a pedantic verification by induction is rather tedious. If now

$$V = \sum_{\varkappa} V_{\varkappa} \subset \mathfrak{C}(n) \qquad (5)$$

is some fixed "domain", namely, a set of n-dimensional elements of the complex $\mathfrak{C}(n)$ (it is always supposed to be finite unless otherwise specified), then the relation

$$(\omega, \chi)_V = \langle V, \omega \smile * \chi \rangle \qquad (6)$$

gives a correct definition of the inner product for forms of the same degree $p = 0, 1, \dots, n$ and allows us to speak of the Hilbert spaces $H^p(V)$. As before, for the forms of different degrees the product (6) is set equal to zero.

1.3. Subdivisions and the limit space. The content of this subsection is closely connected with that of 2.2, Ch. 1. As was pointed out in the Introduction, we shall consider a limiting process in Sec. 3 which connects discrete and continual objects. In that section we shall suppose that a discrete structure appears when a certain domain of the Euclidean space is "subdivided". However, there exists a limiting process of a different type, which makes it possible to connect a certain **topological** space with an appropriately defined sequence of complexes, the combinatorial subdivisions of the initial complex. If, at the same time, the so-called **chain maps** are connected with these subdivisions, then the homology groups of the refined complex automatically turn out to be isomorphic to the groups of the initial complex. These constructions play an important part in defining homology groups of topological spaces and in the proof of their topological invariance.

To illustrate what we have said, we shall consider a trivial one-dimensional example, separating the description of the limiting process from the definition of chain maps. The article [2] can serve as a convenient introduction to the range of problems that have been mentioned and does not require any preliminary information.

Suppose that $\{\mathfrak{C}_{(\nu)}\}$, $\nu = 0, 1, \ldots$ is an infinite sequence of one-dimensional complexes with the generators of the groups of chains (one-dimensional and zero-dimensional) that can be written as $\{e_1; x_1, x_2\}$, $\{e_1, e_2; x_1, x_2, x_3\}, \ldots$, $\{e_1, \ldots, e_{2^\nu}; x_1, x_2, \ldots, x_{2^\nu+1}\}$, and the boundary operator $\partial e_k = x_{\tau k} - x_k$, $\partial x_k = 0$. It is natural to consider the complex $\mathfrak{C}_{(\nu+1)}$ to be the "subdivision" of $\mathfrak{C}_\nu$ resulting from the division of every element e_k into e_{2k-1}, e_{2k} by the "point" x_{2k}. Let us determine the projections $\pi_\nu : \mathfrak{C}_{(\nu+1)} \to \mathfrak{C}_{(\nu)}$, i.e., the mappings of the elements

$$\pi_\nu : e_{2k-1}, e_{2k}, x_{2k} \mapsto e_k, \quad x_{2k-1} \mapsto x_k, \quad x_{2k+1} \mapsto x_{k+1}. \tag{7}$$

REMARK. The projections we have introduced are not **consistent** with the homological structure, i.e., do not commute with the operator ∂. Below I shall indicate a different type of mapping, free from this drawback.

Let us form a topological space E whose points are the sequences

$$\xi = (t_0, t_1, \ldots, t_\nu, \ldots), \qquad t_\nu \in \mathfrak{C}_{(\nu)} \tag{8}$$

(t_ν is one of the generators of e_j, x_j; we do not deal with chains), in which the requirement $\pi_\nu t_\nu = t_{\nu-1}$ is met. Note that every complex possesses a natural partial ordering: $x < y$ if $x \in \partial y$. In particular, $x_{\tau k} < e_k$ in $\mathfrak{C}_{(\nu)}$. We define the collection of the *neighborhoods* (open) of the point $\xi^{(\nu)}$ assuming that for every ν the neighborhood $\mathfrak{U}_\nu(\xi)$ consists of all points $\xi' = (t'_0, t'_1, \ldots, t'_\nu, \ldots)$, for which $t'_\nu \geq t_\nu$.

Sequence (8) is *minimal* if there does not exist a point ξ', different from ξ, such that $t'_\nu \leq t_\nu$ for all ν. We denote the set of all points defined by minimal sequences ("the lower limit of an elementary spectrum" [2]) by $E_0 \subset E$.

THEOREM. *The set E_0 is homeomorphic to the closed interval l of the real axis.*

The statement of the theorem follows from the existence of a one-to-one correspondence between the minimal sequences and the points l regarded as the limits of the chains of imbedded intervals resulting from the successive divisions of l into 2^ν parts. At the same time, we can establish a one-to-one correspondence between the neighborhoods in E_0 and in l, and this ensures their homeomorphism [2]. ■

Let us consider now the correspondence between the complex $\mathfrak{C}_{(\nu)}$ and $\mathfrak{C}_{(\nu+1)}$ (the "subdivision" of $\mathfrak{C}_{(\nu)}$), realized by *chain mappings*, i.e., mappings of chains which commute with the operation ∂. As was noted, a correspondence of this kind between complexes ensures an isomorphism of homology groups.

We preserve the notations used in (7) and note by additional primes the generators of the complex $\mathfrak{C}_{(\nu+1)}$ (as distinct from the generators of the complex $\mathfrak{C}_{(\nu)}$). We define the injection I and the projection π

$$I: \ \mathfrak{C}_{(\nu)} \to \mathfrak{C}_{(\nu+1)}, \qquad \pi: \ \mathfrak{C}_{(\nu+1)} \to \mathfrak{C}_\nu,$$

setting

$$Ix_k = x'_{2k-1}, \qquad Ie_k = e'_{2k-1} + e'_{2k}, \qquad I\cdot 0 = \pi \cdot 0 = 0,$$

$$\pi x'_{2k-1} = \pi x'_{2k} = x_k, \qquad \pi e'_{2k-1} = 0, \qquad \pi e'_{2k} = e_{2k}.$$

The relation $\pi\partial = \partial\pi$, $I\partial = \partial I$ can be immediately verified (for instance, $\pi\partial e'_{2k-1} = \pi(x'_{2k} - x'_{2k-1}) = x_k - x_k = 0 = \partial\pi e'_{2k-1}$). At the same time we note that $\pi I = 1$, $I\pi \neq 1$.

The general definition of a subdivision of a complex and the concept of a subcomplex, which is closely connected with it, can be found, for instance, in [30, p. 220, 21, p. 30]. The constructions similar to those we have carried out can be realized in the case of simplicial (or cubic) complexes of arbitrary dimensions.

2. Difference Operators and the Principal Problems

2.0. Preliminary remarks. The definitions of the complex $K(n)$, the operator d, and the inner product introduced in Sec. 1 allow us to define the operator δ. Subsection 2.1 is devoted to the techniques of calculation and the notation of discrete (difference) d, δ. Subsections 2.2 and 2.3 contain discussions of the discrete analogs of problems that we studied in Ch. 2. The main distinction, besides discreteness, is a direct (without the use of a double) inclusion into consideration of homogeneous boundary conditions in a domain with boundary. In this section we shall mainly consider general schemes. Secs. 3 and 4 contain a more thorough investigation of some concrete situations.

2.1. Difference operators d, δ. We begin with the remark that in a multidimensional case, as well, the relationship between the operator d with difference operators can be formally established in accordance with the scheme given in Sec. 1, Ch. 0, applied to the construction considered in 1.2. However, this way is rather complicated. It is easy to verify that an equivalent result can be obtained by a direct use of the analogy with a continual case.

Thus, if, for $n = 3$, $\omega = (u, v, w)$ is a 1-form written as the sum

$$\omega = \sum_{\varkappa} (u_\varkappa \varepsilon_1^\varkappa + v_\varkappa \varepsilon_2^\varkappa + w_\varkappa \varepsilon_3^\varkappa),$$

where the subscript $\varkappa = (\varkappa_1, \varkappa_2, \varkappa_3)$ is the indicator of the zero-dimensional element of our complex-model, and $\varepsilon_p^\varkappa$, $p = 1, 2, 3$ are one-dimensional base elements mentioned in the remark in 1.2, then the representation for $d\omega$ can be written in the form

$$d\omega = \sum_{\varkappa} \{(\Delta_1 v_\varkappa - \Delta_2 u_\varkappa)\varepsilon_{12}^\varkappa + (\Delta_1 w_\varkappa - \Delta_3 u_\varkappa)\varepsilon_{13}^\varkappa +$$

$$(\Delta_2 w_\varkappa - \Delta_3 v_\varkappa)\varepsilon_{23}^\varkappa\}, \tag{1}$$

where Δ_p are partial differences ($\Delta_1 v_\varkappa = v_{\tau\varkappa_1\varkappa_2\varkappa_3} - v_{\varkappa_1\varkappa_2\varkappa_3}$, etc.) and $\varepsilon^\varkappa_{pq}$ are the corresponding two-dimensional base elements. The connection with the relation

$$d\omega = (D_1 v - D_2 u)dxdy + (D_1 w - D_3 u)dxdz + (D_2 w - D_3 v)dydz$$

is obvious.

In this section we shall not use the "explicit" form of the operator d (or of the operator δ defined below). We shall need this form of notation only in 3.3 and 3.4. For the time being, it is essential to consider more attentively the discrete analog of the Green formula that relates the operators d and δ (cf. Ch. 2).

Having fixed the domain V of the form (5), Sec. 1 (we shall write the term "domain" without quotation marks) and using the formalism of 1.2, we can write, for the forms ω, χ of the degrees $p-1$ and p, respectively, the relation

$$(d\omega, \chi)_V = \langle V, d\omega \smile *\chi \rangle = \langle V, d\{\omega \smile *\chi\}\rangle - Q(\omega)\langle V, \omega \smile d*\chi\rangle =$$

$$\langle \partial V, \omega \smile *\chi \rangle + (\omega, \delta\chi)_V, \tag{2}$$

that defines the operator δ:

$$\delta\chi = -Q(\omega) *^{-1} d * \chi, \qquad Q(\omega) = (-1)^{\deg \omega}. \tag{3}$$

If the "boundary terms", namely, the first term of the last row of (2), vanish, we obtain

$$(d\omega, \chi)_V = (\omega, \delta\chi)_V$$

These conditions can be chosen in the form of the requirement

$$\omega|(\partial V)_{p-1} = 0 \tag{Γ}$$

or

$$*\chi|(\partial V)_{n-p} = 0, \tag{Γ^*}$$

where $(\partial V)_{p-1}$, $(\partial V)_{n-p}$ are collections of "boundary elements" of the domain V of the corresponding dimension.

An unfortunate property of relation (2), which is of fundamental importance for the subsequent considerations, is the inadequacy mentioned in Sec. 1, Ch. 0 of defining ω, χ only as elements of the spaces $H^{p-1}(V)$, $H^p(V)$, respectively, when this relation is used. Relation (2) includes "superfluous" components of the forms ω, χ. The content of Sec. 3 illustrates the consequences of this phenomenon which must be taken into account when concrete problems are analyzed. For the time being, we shall only take into consideration the remarks that refer to the formal side of the problem which has two aspects, one of which is connected with our method of definition of an inner product and the spaces H^p and the other is connected with the "nonlocalization" of the discrete operators d, δ.

Let us consider the first aspect. We shall introduce some auxiliary concepts. The collection $\mathfrak{C}'$ of elements from $\mathfrak{C}(n)$ forms a *closed subcomplex* $\mathfrak{C}' \subset \mathfrak{C}(n)$ if every element $s_{(p)} \in \mathfrak{C}'$ enters into $\mathfrak{C}'$ together with its boundary, i.e., $\partial s_{(p)} \in \mathfrak{C}'$. The collection of p-dimensional elements of $\mathfrak{C}'$ is known as its *p-skeleton.*

In particular, we can associate the closed subcomplex $\overline{V}_\varkappa$ with every n-element

$$V_\varkappa = e_{\varkappa_1} \otimes \ldots \otimes e_{\varkappa_n} \in \mathfrak{C}(n),$$

including into $\overline{V}_\varkappa$ $2n$ elements $s^k_{(n-1)}$ generating $\partial V_\varkappa$, $n(n-1)$ elements entering into the boundaries of $s^k_{(n-1)}$ and so on. If we now assume that V in (5), Sec. 1, consists of a single element $V_\varkappa$ and $\omega = \{\omega_{\varkappa_1 \ldots \varkappa_p}\}$ is a form of degree p, then the inner product $(\omega, \omega)_{V_\varkappa}$ will include C^n_p components of ω corresponding to the "point" $\varkappa$ whereas the p-skeleton of the subcomplex $\overline{V}_\varkappa$ contains $2C^n_p$ elements and, in particular, all of them enter into the p-skeleton of the boundary ∂V; and this is essential if we want to use (2). We have to take into account this phenomenon in domain (5), Sec. 1, of the general form, in the neighborhood of a certain part of the boundary ∂V.

As concerns the second aspect, it is sufficient to note that it is obviously necessary to define the forms under consideration on the skeleton of the complex $\mathfrak{U}\overline{V}$, which is wider than $\overline{V}$, in order to define $d\omega$ (or $\delta\omega$) on certain elements which belong to ∂V (or are close to ∂V).

As to the "explicit" notation of the operator δ defined by (3) (the notation similar to (1) for d), it is most convenient to use, just as for d, the analogy with a continual definition. However, in this case the partial differences have a somewhat different form. For example, the analog of the partial derivative $-D_2 v$, entering into δv, is the difference

$$\Delta_2^\sigma v_\varkappa = v_{\varkappa_1 \sigma \varkappa_2 \varkappa_3} - v_{\varkappa_1 \varkappa_2 \varkappa_3}.$$

2.2. Orthogonal decompositions and cohomology. We fix the dimension n for $\mathfrak{C}(n)$ and the domain V defined by the finite sum (5), Sec. 1. We write H^p instead of $H^p(V)$. We understand K^p to be a linear space of forms of degree p defined over the whole p-skeleton $\mathfrak{C}^p \subset \mathfrak{C} \equiv \mathfrak{C}(n)$. Let $\mathcal{D}_\delta^0 \subset K^{p+1}$ be a linear space of forms subject additionally to homogeneous conditions (Γ^*), 2.1; by $\mathfrak{R}_\delta^0 \subset H^p$ we denote the range of the operator δ defined on $\mathcal{D}_\delta^0$. Since $\mathfrak{R}_\delta^0$ is obviously a subspace of H^p, we can write the orthogonal decomposition

$$H^p = \mathfrak{R}_\delta^0 \oplus \mathfrak{N}_d,$$

where, as follows from (2), the subspace $\mathfrak{N}_d$ consists of the elements $\varphi \in H^p$ that satisfy the relation $(\delta\chi, \varphi) = 0$ for any $\chi \in \mathcal{D}_\delta^0$.

We denote by $\mathfrak{R}_d \subset H^p$ the subspace of the elements φ that can be represented as $\varphi = d\omega$, $\omega \in K^{p-1}$.

PROPOSITION 1. *The inclusion*

$$\mathfrak{R}_d \subset \mathfrak{N}_d$$

holds true.

This statement follows directly from (2) and from the definitions we have introduced. ■

Using now the decomposition $\mathfrak{N}_d = \mathfrak{R}_d \oplus \mathcal{H}_d$, we can write

$$H^p = \mathfrak{R}_\delta^0 \oplus \mathfrak{R}_d \oplus \mathcal{H}_d. \tag{4}$$

REMARK. The subspace $\mathcal{H}_d$ is obviously an analog of the subspace $\mathfrak{N}_d'$ from 2.2, Ch. 2, but now the relationship with the homology theory is even more evident. We must now include the index, which indicates the boundary conditions, into the notations and make slight changes in the notation in order to make it more convenient.

Considering, at the same time, the orthogonal complement of $\mathfrak{R}_d$ in H^p, we can represent H^p in the form

$$H^p = \mathfrak{R}_d \oplus \mathfrak{N}_\delta^0.$$

PROPOSITION 2. *The inclusion* $\mathfrak{R}_d^0 \subset \mathfrak{N}_\delta^0$ *holds true.* ■

Representing now $\mathfrak{N}_d^0$ in the form $\mathfrak{R}_d^0 \oplus \mathcal{H}_\delta$, writing $H^p = \mathfrak{R}_d \oplus \mathfrak{R}_d^0 \oplus \mathcal{H}$, and comparing it to (4), we have

$$\mathcal{H}_\delta = \mathcal{H}_d = \mathfrak{N}_\delta^0 \cap \mathfrak{N}_d = \mathcal{H}.$$

The resulting decomposition

$$H^p = \mathfrak{R}_d \oplus \mathfrak{R}_\delta^0 \oplus \mathcal{H}$$

is an analog of the **Kodaira theorem** for a manifold with boundary. In this case, the subspace $\mathcal{H}$ plays the part of the homology group for V.

These constructions admit of many variants. For example, beginning (instead of $\mathcal{D}_\delta^0$) with $\mathcal{D}_d^0 \subset K^{p-1}$, i.e., with the subspace of forms subject to the condition (Γ) from 2.1, and reasoning as above, we find, instead of (4),

$$H^p = \mathfrak{R}_d^0 \oplus \mathfrak{R}_\delta \oplus \tilde{\mathcal{H}}_\delta,$$

where $\tilde{\mathcal{H}}_\delta$ is a subspace of "δ-homologies" for V. As a rule, the subspaces $\mathcal{H}$, $\tilde{\mathcal{H}}_\delta$ turn out to be isomorphic. Another variant is the isolation of the subspace of forms which are "harmonic inside V":

$$H^p = \mathfrak{R}_d^0 \oplus \mathfrak{R}_\delta^0 \oplus \mathfrak{N}_\Delta$$

(cf. Sec. 3).

2.3. Natural equations. It is sometimes convenient to say that the equations (systems) considered in Sec. 3, Ch. 2, are "natural", thereby emphasizing their invariant notation by means of d, δ. Now this notation is complemented by the "natural" discrete analogs.

Continuing to consider a certain domain V to be fixed, we shall discuss in detail the case $\dim V = 3$. We write

$$\begin{aligned} d\omega + \delta\varphi &= f \\ \delta\omega &= h, \end{aligned} \qquad (L)$$

$$\begin{aligned} d\psi + \delta\chi &= g \\ d\chi &= k, \end{aligned} \qquad (L^t)$$

where φ, χ, ω, ψ are forms of degrees 3, 2, 1, 0, and the degrees of the right-hand sides are defined correspondingly. If we introduce, just as in Sec. 3, Ch. 2, symbolic notations $(\varphi, \omega) = \omega_{\mathrm{I}}$, $(\chi, \psi) = \chi_{\mathrm{II}}$ (the collections of forms of an odd and an even degree), then

$$(L\omega_{\mathrm{I}}, \chi_{\mathrm{II}})_V = \langle \partial V, \omega \smile *\chi - \chi \smile *\varphi - \psi \smile *\omega \rangle - (\omega_{\mathrm{I}}, L^t \chi_{\mathrm{II}})_V \quad (5)$$

is the Green formula that connects the formally conjugate systems (L), (L^t).

REMARK. Relation (5) preserves its sense in a continual case (when the pairing $\langle \partial V, \ldots \rangle$ is replaced by a surface integral), and, as was noted in Sec. 4, Ch. 2, can be used for investigating the corresponding differential equations.

An ordinary way of using (5) for studying boundary value problems is the introduction into consideration of **conjugate** homogeneous boundary conditions which ensure the vanishing of the boundary terms. We take, as these conditions, the requirements

$$* \chi | (\partial V)^1 = 0 \quad \text{or} \quad * \omega | (\partial V)^2 = * \varphi | (\partial V)^0 = 0, \tag{6}$$

where $(\partial V)^p$ is a p-dimensional skeleton of the boundary. When (6) is satisfied, the operators (L), (L^t) turn out to be conjugate, i.e., satisfying the relation

$$(L\omega_{\rm I}, \chi_{\rm II})_V = (\omega_{\rm I}, L^t \chi_{\rm II})_V.$$

If $\chi_{\rm II}$ is a solution of the homogeneous system (L^t), this solution satisfying conditions (6), then, using decomposition (4) for H^1, we can infer that

$$d\chi = \delta\chi = 0, \qquad d\psi = 0,$$

i.e., $\chi \in \mathcal{H}^2$, $\psi \in \mathcal{H}^0$. It follows, in the domain V "homeomorphic to a ball", that $\chi = 0$ and ψ is defined with an accuracy to within a constant.

Similar arguments can also be applied to system (L). The conjugacy of L, L^t allows us to infer that the problem being considered is solvable for (L^t) for any right-hand side and for (L) under the condition of orthogonality of h to constants.

Returning to the remark concerning the continual analog for (5), we must note that (6) corresponds either to the specification of the covector $* \chi$ throughout the boundary ∂V or to the specification of the scalar $* \varphi$ on the boundary and a normal component of the covector ω. The asymmetry among the conditions is only illusory. Thus, in a discrete case, for a "parallelepiped", the number of elements appearing in $(\partial V)^1$ coincides with that in $(\partial V)^2 \cup (\partial V)^0$.

The unique solvability of the problems we have considered for (L), (L^t) and the finite-dimensionality of the Hilbert spaces H^p used make it possible to obtain a number of relations which are typical of the continual theory of the objects under consideration. It follows from our constructions that for the solutions of system (L^t), subject to conditions (6), we have

$$\|\chi_{\mathrm{II}}\|^2 \equiv (\chi_{\mathrm{II}}, \chi_{\mathrm{II}})_V \le c\|g_{\mathrm{I}}\|^2 + \|\chi_{\mathcal{H}}\|^2, \tag{7}$$

where $g_{\mathrm{I}} = (g, k)$ and $\chi_{\mathcal{H}}$ is the projection of χ onto $\mathcal{H}^2$. At the same time,

$$\|g_{\mathrm{I}}\|^2 = (L^t\chi_{\mathrm{II}}, L^t\chi_{\mathrm{II}})_V =$$

$$(d\psi, d\psi)_V + (\delta\chi, \delta\chi)_V + 2(d\psi, \delta\chi)_V + (d\chi, d\chi)_V, \tag{8}$$

and

$$(d\psi, \delta\chi)_V = -\langle \partial V, d\psi \smile *\chi\rangle + (dd\psi, \chi)_V = 0$$

by virtue of (6) and the properties of the operation d. Now we infer from (7), (8) that

$$(\chi_{\mathrm{II}}, \chi_{\mathrm{II}})_V \le c\{(d\chi_{\mathrm{II}}, d\chi_{\mathrm{II}})_V + (\delta\chi_{\mathrm{II}}, \delta\chi_{\mathrm{II}})_V\} + \|\chi_{\mathcal{H}}\|^2. \tag{9}$$

Similar constructions can be realized for solving the system (L) and can be extended to the general systems (L), (L^t), Sec. 3, Ch. 2, corresponding to the n-dimensional case.

REMARK. It stands to reason that the technique for obtaining (7), (9) that we have used cannot be used when we introduce a scale and "refine" the complex $\mathfrak{C}(n)$ which increases indefinitely the dimension (finite) of the space $H^p(V)$, the "geometric" domain V remaining fixed. The corresponding procedures of subdivision are necessary for considering the approximations of continual objects by discrete ones. The constructions suitable in this case are given in Sec. 3.

Using this remark on the validity of the analogs of inequalities (7), (9) in the n-dimensional case, we shall formulate the corollaries that refer to the Poisson equation

$$-\Delta\chi \equiv (d\delta + \delta d)\chi = f, \qquad \chi, f \in H^p. \tag{10}$$

PROPOSITION 3. *For any f orthogonal to $\mathcal{H}^p$, there exists a unique solution of Eq.* (10), *orthogonal to $\mathcal{H}^p$, which satisfies the conditions*

$$*\chi|(\partial V)^{n-p} = \chi|(\partial V)^p = 0. \tag{11}$$

PROOF. Let us consider the inner product $(\Delta\chi, \chi)_V$. Since

$$\begin{aligned}(d\delta\chi, \chi)_V &= \langle \partial V, \delta\chi \smile *\chi\rangle + (\delta\chi, \delta\chi)_V,\\ (\delta d\chi, \chi)_V &= \langle \partial V, -\chi \smile *d\chi\rangle + (d\chi, d\chi)_V\end{aligned} \tag{12}$$

and conditions (11) ensure the vanishing of the boundary terms, the kernel of the operator Δ belongs to $\mathcal{H}$. At the same time, it follows from (12) that the relation $(\Delta\chi, \omega)_V = (\chi, \Delta\omega)_V$ is valid for χ, ω satisfying (11), i.e., the problem is self-conjugate. ■

It pays to note that $(\partial V)^n = 0$ and the first of conditions (11) is absent for $p = 0$; similarly, the second condition of (11) is absent for $p = n$.

In conclusion, we shall consider an example of natural equations of a different kind, namely, the following overdetermined system.

PROPOSITION 4. *The equations*

$$d\chi = f, \quad \delta\chi = h, \qquad \chi \in H^p,$$

are always solvable when the compatibility conditions

$$df = 0, \quad f \perp \mathcal{H}^{p+1}, \qquad \delta g = 0, \quad g \perp \tilde{\mathcal{H}}^{p-1},$$

are fulfilled. The solutions are defined with an accuracy to within an arbitrary term from the subspace $\mathfrak{N}_d \cap \mathfrak{N}_\delta$.

We invite the reader to carry out the verification that uses the orthogonal decompositions of form (4). ■

A variant of Proposition 4, that contains certain boundary conditions for χ, is also possible. It is also clear that in the range of ideas we have discussed we can consider the Maxwell equations, the analogs of regular systems from Ch. 2, a time-including equation, and other discrete models of natural equations.

3. A Two-Dimensional Case: Boundary Value Problems and Approximations

3.0. Preliminary remarks. A characteristic feature of the study of difference equations given below is the use of special formalism that makes it possible to analyze in detail the purely combinatorial relations that are encountered. The construction of this formalism is stimulated by the above-mentioned far-reaching analogy with continual objects. It is natural to require that the next stage of the consideration be the revealing of the actual connection between the combinatorial objects having been constructed and continual objects, i.e., the revealing of the possibility of the approximation and the limiting process.

In this section, we make more clear (for a rectangle, which is the simplest two-dimensional domain), on one hand, the combinatorial considerations, which were somewhat rough in Sec. 2; and, on the other hand, describe the methods of associating the corresponding functions and differential relations with discrete "forms" and equations, which have not yet been touched upon (except for Ch. 0). These methods differ slightly from the ordinary ones and are mainly based on the use of the simplest splines resulting from the averaging of step functions. One of the formal results is the construction of nonstandard approximations of the generalized solutions of the Poisson equation under the minimal requirements of smoothness of the right-hand side ($f \in \mathbb{H}(V)$).

3.1. Continual objects. Here we shall consider, in particular, the approximation of the solutions of the equations

$$-D_y u + D_x v = f, \qquad D_x u + D_y v = 0 \tag{1}$$

in the rectangle $\Omega \subset \mathbb{R}^2$ under the condition

$$u \cos nx + v \cos ny = 0 \tag{2}$$

imposed on $\partial\Omega$. In this event, the functions u, v play the part of the components of the 1-form

$$\omega(x, y) = u\, dx + v\, dy.$$

At the same time, we shall have to deal with the scalars $\varphi(x, y)$ and the 2-forms $\eta(x, y)\, dx\, dy$.

With the use of the operators d, δ we can write Eqs. (1) as

$$d\omega = f, \qquad -\delta\omega = 0. \tag{3}$$

In conjunction with (3), the related equations

$$d\varphi + \delta\eta = g, \tag{4}$$

$$-\Delta\xi \equiv (d\delta + \delta d)\xi = h \tag{5}$$

take part in the constructions as well as the orthogonal decompositions of the Hilbert space $\mathbb{H}^1(\Omega)$ of the 1-forms,

$$\mathbb{H}^1 = \mathfrak{R}_d^0 \oplus \mathfrak{R}_\delta \oplus I, \qquad \mathbb{H}^1 = \mathfrak{R}_d^0 \oplus \mathfrak{R}_\delta^0 \oplus \mathfrak{N}_\Delta, \tag{6}$$

which are continual variants of decomposition (4), Sec. 2. We shall consider the Dirichlet problem for Eqs. (5).

In addition to the space $\mathbb{H}^1$ it is useful to employ the Hilbert spaces $W^1(\Omega)$, $\dot{W}^1(\Omega)$ of functions with the generalized first derivatives and the norm

$$|\xi, W^1|^2 = (d\xi, d\xi) + (\delta\xi, \delta\xi) + (\xi, \xi)$$

(cf. 5.2, Ch. 1). In $\dot{W}^1$ the element ξ is subjected to the homogeneous boundary conditions, and the term (ξ, ξ) is usually absent in the definition of the norm.

Note that it would be interesting to consider, in the context of this section, constructions connected with the Cauchy integral (5.3, Ch. 2).

3.2. Combinatorial structures. We shall use, in a two-dimensional case, the model of the Euclidean space, namely, the complex (1.2, Sec. 1) and introduce special notations for its elements. We set

$$x_k \otimes x_s = x_{k,s}, \qquad e_k \otimes x_s = e^1_{k,s},$$

$$e_k \otimes e_s = V_{k,s}, \qquad x_k \otimes e_s = e^2_{k,s}.$$

Correspondingly, we have

$$\partial x_{k,s} = 0, \qquad \partial e^1_{k,s} = x_{\tau k,s} - x_{k,s}, \qquad \partial e^2_{k,s} = x_{k,\tau s} - x_{k,s},$$

$$\partial V_{k,s} = e^1_{k,s} + e^2_{\tau k,s} - e^2_{k,s} - e^1_{k,\tau s}.$$

Similar notations can be obviously introduced for the base elements of the conjugate complex $K(2)$, but we shall not use them but shall immediately pass to systems of coefficients that define the forms

$$\varphi = \{\varphi_{k,s}\}, \qquad \omega = \{u_{k,s}, v_{k,s}\}, \qquad \eta = \{\eta_{k,s}\}. \tag{7}$$

We regard forms (7) as defining the corresponding functions over $\mathfrak{C}(2)$:

$$\varphi | x_{k,s} = \varphi_{k,s}, \qquad \omega | e^1_{k,s} = u_{k,s},$$

$$\omega|e^2_{k,s} = v_{k,s}, \qquad \eta|V_{k,s} = \eta_{k,s}.$$

We give the expression for the operator d and simultaneously introduce special notations for the differences

$$\begin{aligned} d\varphi|e^1_{k,s} &= \varphi_{\tau k,s} - \varphi_{k,s} \equiv \Delta_k\varphi_{k,s}, \\ d\varphi|e^2_{k,s} &= \varphi_{k,\tau s} - \varphi_{k,s} \equiv \Delta_s\varphi_{k,s}, \\ d\omega|V_{k,s} &= v_{\tau k,s} - v_{k,s} - u_{k,\tau s} + u_{k,s} = \Delta_k v_{k,s} - \Delta_s u_{k,s}, \end{aligned} \tag{8}$$

that will be convenient for the subsequent considerations. We define the principal domain, i.e., the rectangle V, as the sum

$$V = \sum_{k,s} V_{k,s}, \qquad k = 1, \ldots, N, \quad s = 1, \ldots, M. \tag{9}$$

We agree that in what follows, unless the limits of summation are specified, the subscripts k, s always run over the set of values indicated in (9).

The inner product of forms of the same degree, defined in Sec. 1, reduces now to the componentwise sums

$$(\varphi, \psi)_V = \sum_{k,s} \varphi_{k,s}\psi_{k,s},$$

and the same refers to $(\omega, \vartheta)_V$, $(\eta, \chi)_V$. The corresponding Hilbert space H^p, $p = 0, 1, 2$, are of dimension MN and $2MN$, respectively.

Here is the expanded notation of the Green formula (2) from 2.1 for the pair $\varphi \in K^1$, $\omega \equiv K^2$.

PROPOSITION 1. *The identity*

$$(d\varphi, \omega)_V \equiv \sum_{k,s}(\Delta_k\varphi_{k,s} \times u_{k,s} + \Delta_s\varphi_{k,s} \times v_{k,s}) =$$

$$\sum_k(\varphi_{k,\tau M}v_{k,M} - \varphi_{k,1}v_{k,0}) + \sum_s(\varphi_{\tau N,s}u_{N,s} - \varphi_{1,s}u_{0,s}) +$$

$$\sum_{k,s}\varphi_{k,s}(-\Delta_k u_{\sigma k,s} - \Delta_s v_{k,\sigma s}) \tag{10}$$

holds true.

In accordance with notations (8), we have $\Delta_k u_{\sigma k,s} = u_{k,s} - u_{\sigma k,s}$ and a similar representation can be used for $\Delta_s v_{k,\sigma s}$. The direct verification of identity (10) is elementary although somewhat tedious. ■

It was repeatedly emphasized that (10) included not only the components $\varphi_{k,s}$, $u_{k,s}$, $v_{k,s}$, that define the corresponding elements of the spaces $H^0(V)$, $H^1(V)$ (in this case the subscripts k, s would run only over the values from (9)), but also the components $\varphi_{k,\tau M}$, $v_{k,0}$, etc.

Identity (10) obviously gives the expression for the operation $\delta\omega$:

$$\delta\omega|x_{k,s} = -\Delta_k u_{\sigma k,s} - \Delta_s v_{k,\sigma s}.$$

Correspondingly, the last term in (10) can be written as $(\varphi, \delta\omega)_V$.

Alongside (10), it is convenient to have an explicit notation of a similar relation for the pair $\omega \equiv K^1$, $\eta \in K^2$.

PROPOSITION 2. *The identity*

$$(d\omega, \eta)_V \equiv \sum_{k,s}(\Delta_k v_{k,s} - \Delta_s u_{k,s}) =$$

$$\sum_k (u_{k,1}\eta_{k,0} - u_{k,\tau M}\eta_{k,M}) + \sum_s (v_{\tau N,s}\eta_{N,s} - v_{1,s}\eta_{0,s}) +$$

$$\sum_{k,s}(u_{k,s}\Delta_s\eta_{k,\sigma s} - v_{k,s}\Delta_k\eta_{\sigma k,s}) \tag{11}$$

holds true. ■

Now we have for $\delta\eta$ the relations

$$\delta\eta|e^1_{h,s} = \Delta_s\eta_{k,\sigma s}, \qquad \delta\eta|e^2_{h,s} = -\Delta_k\eta_{\sigma k,s}$$

and we can write the last term in (11) as $(\omega, \delta\eta)_V$.

From the point of view of arithmetic, identities (10), (11) are equivalent and can be obtained one from the other by a change in the notations.

As was pointed out, relation (2) from Sec. 1, in its "explicit" notation gives Abel transform in a one-dimensional case. Thus we can regard (10), (11) as its two-dimensional analogs.

3.3. Equations and problems. We begin with the discrete Dirichlet problem for the Poisson equation. In this equation, written as relation (5), the form ξ can be of degree 0, 1, or 2. We shall consider in detail the classical case

$$-\Delta\varphi \equiv \delta d\psi = g, \tag{12}$$

where φ, g are 0-forms. Having written (12) in a "pointwise" form

$$\delta d\,\varphi|x_{k,s} \equiv 4\varphi_{k,s} - \varphi_{\sigma k,s} - \varphi_{k,\sigma s} - \psi_{k,\tau s} = g_{k,s}, \tag{13}$$

we see that we have a conventional difference analog of the Laplace operator, but without a "scale" and normalization of the differences. The following statement plays a significant part in the sequel.

PROPOSITION 3. *The identity*

$$(d\varphi, d\psi)_V \equiv \sum_{k,s}(\Delta_k\varphi_{k,s}\Delta_k\psi_{k,s} + \Delta_s\varphi_{k,s}\Delta_s\psi_{k,s}) =$$

$$\sum_k[\varphi_{k,\tau M}(\psi_{k,\tau M} - \psi_{k,M}) - \varphi_{k,1}(\psi_{k,1} - \psi_{k,0})] +$$

$$\sum_s[\varphi_{\tau N,s}(\psi_{\tau N,s} - \psi_{N,s}) - \varphi_{1,s}(\psi_{1,s} - \psi_{0,s})] + (\varphi, \delta\, d\psi)_V. \tag{14}$$

holds true.

This identity results from (10) upon the substitution of $d\psi$ for ω. ∎

DEFINITION. We call the form $\varphi \in H^0$ the *solution of the Dirichlet problem* for Eq. (12) if relations (13), in which

$$\varphi_{k,0} = \varphi_{k,\tau M} = 0, \qquad \varphi_{0,s} = \varphi_{\tau N,s} = 0, \tag{15}$$

are satisfied for all $k = 1, \dots, N$, $s = 1, \dots, M$.

Conditions (15) make it possible to consider the operator $-\Delta$ (corresponding to the Dirichlet problem) to be defined on the arbitrary element $\varphi \equiv H^0$ (conditions (15) give the necessary extension of the definition of φ "beyond H^0"). In other words, the definition we have introduced is given by the operator

$$-\Delta:\ H^0 \to H^0. \tag{16}$$

REMARK. The figure in parenthesis, added to the number of a lemma or a theorem, indicates to which of the spaces H^p, $p = 0, 1, 2$, the result refers.

LEMMA 1 (0). *Operator* (16) *defined in the way described above is self-conjugate.*

PROOF. We suppose that the form ψ in (14) is also subject to conditions of form (15) and obtain

$$(d\varphi, d\psi)_V = -\sum_k \varphi_{k,1}\psi_{k,1} - \sum_s \varphi_{1,s}\psi_{1,s} + (\varphi, \delta\, d\psi)_V. \tag{17}$$

It follows that under the indicated conditions,

$$(\delta\, d\varphi, \psi)_V = (\varphi, \delta\, d\psi)_V. \ \blacksquare$$

THEOREM 1 (0). *For any $g \in H^0$ a solution of the Dirichlet problem for Eq.* (12) *exists and is unique.*

PROOF. By virtue of the self-conjugacy of the operator which corresponds to the problem, it is sufficient to establish the uniqueness of the solution. Setting $g = 0$ and using (14), (17) on the assumption that $\psi = \varphi$, we obtain

$$\sum_{k,s}[(\Delta_k\varphi_{k,s})^2 + (\Delta_s\varphi_{k,s})^2] + \sum_k \varphi_{k,1}^2 + \sum_s \varphi_{1,s}^1 = 0,$$

whence it follows that $\varphi = 0$. $\blacksquare$

Passing to the equations

$$-\Delta\eta \equiv d\delta\eta = f, \qquad \eta, f \in H^2, \tag{18}$$

$$-\Delta\omega \equiv (d\delta + \delta d)\omega = r, \qquad \omega, r \in H^1, \tag{19}$$

we note that the consideration of (18) does not differ from that carried out for the 0-forms φ. The analog of relation (14) results from (11) upon the substitution $\omega = \delta\chi$, and this gives the notation of the inner product $(\delta\chi, \delta\eta)_V$. Its properties are completely similar to those of $(d\varphi, d\psi)_V$.

The situation with relation (19), which corresponds to a pair of Poisson "scalar" equations, is more difficult. Here we must study the sum $(d\omega, d\vartheta)_V + (\delta\omega, \delta\vartheta)_V$. We shall consider it later, in the case of $\vartheta = \omega$, in connection with the object which is of special interest to us, namely, Eq. (1) (or (3)). For the time being we shall restrict the discussion of (18), (19) by the remarks we have made and with the formulation of the result.

DEFINITION. We call the element $\eta \in H^2$ $(\omega \in H^1)$ the *solution of the Dirichlet problem* for (18) (for (19)) if relations of form (13) with the corresponding right-hand sides are satisfied for the components $\eta_{k,s}$ (components $u_{k,s}$, $v_{k,s}$) and the homogeneous conditions (15) are taken into account for the numbers of components indicated in (15).

LEMMA 1 (2, 1). *The operators*

$$\Delta : \ H^p \to H^p, \qquad p = 1, 2,$$

defined in the way described above are self-conjugate. ∎

THEOREM 1 (2, 1). *For any right-hand sides from H^p, $p = 1, 2$, the solutions of the Dirichlet problem for Eqs.* (18), (19) *exist and are unique.* ∎

Let us pass to equations of the form (3)

$$d\omega = f, \qquad \delta\omega = g, \tag{20}$$

without supposing that $g = 0$. We shall need the analog of Proposition 3 mentioned above.

PROPOSITION 4. *When the conditions*

$$u_{0,s} = u_{k,\tau M} = 0, \qquad v_{k,0} = v_{\tau M,s} = 0,$$
$$k = 1, \dots, N, \qquad s = 1, \dots, M, \tag{21}$$

are fulfilled, the relation

$$(d\omega, d\omega)_V + (\delta\omega, \delta\omega)_V =$$
$$\sum_{k,s} [(\Delta_k v_{k,s})^2 + (\Delta_s v_{k,\sigma s})^2 + (\Delta_k u_{\sigma k,s})^2 + (\Delta_s u_{k,s})^2] \tag{22}$$

holds true.

PROOF. In a direct computation, the right-hand side must contain, besides the terms that have been written, the sum of doubled products of the form

$$2\sum_{k,s} [(\Delta_k u_{\sigma k,s})(\Delta_s v_{k,\sigma s}) - (\Delta_k v_{k,s})(\Delta_s u_{k,s})].$$

After canceling and grouping of terms, this sum reduces to the "sum of boundary terms":

$$2\sum_k [(\Delta_k u_{\sigma k,\tau M}) v_{k,M} - (\Delta_k u_{\sigma k,s}) v_{k,0}] +$$
$$2\sum_s [(\Delta_s u_{0,s}) v_{1,s} - (\Delta_s u_{N,s}) v_{\tau N,s}], \tag{23}$$

and then the validity of the statement is obvious (the terms with the factor $u_{0,\tau M}$ that are not subject to conditions (21) are canceled out). ∎

DEFINITION. We say that the *solution of Eqs.* (20) *under conditions* (21) is the form $\omega \equiv H^1$ which satisfies the system of relations

$$\begin{aligned} &(\Delta_k v_{k,s} - \Delta_s u_{k,s}) | V_{k,s} = f_{k,s} \qquad &k = 1, \dots, N, \\ &(-\Delta_k u_{\sigma k,s} - \Delta_s v_{k,\sigma s}) | x_{k,s} = g_{k,s}, \quad &s = 1, \dots, M, \end{aligned} \tag{24}$$

in which conditions (21) are taken into account.

THEOREM 2. *The solution of problem* (20)–(21) *defined in the above-indicated way exists and is unique for any right-hand sides* $f \in H^2$, $g \in H^0$.

The uniqueness of the solution immediately follows from Proposition 4. (Setting $f = g = 0$, we obtain $u = v = 0$ from (21), (22).) The existence of the solution for any right-hand sides would be a consequence of the uniqueness of the solution of the conjugate problem defined in an appropriate way. The reader can verify that this uniqueness actually holds true. ■

The absence of a complete proof is not a principal drawback of this theorem. The main drawback is that conditions (21) we have used do not correspond to the boundary conditions (2) which are most frequently used in a continual case (the definition of a normal component of the covector ω). In (21) the component u is defined on the left and upper sides of V and v is defined on the right and lower sides. Conditions (2) would correspond to the conditions

$$u_{0,s} = u_{\tau N,s} = 0, \qquad v_{k,0} = v_{k,\tau M} = 0, \tag{25}$$

but they do not serve for vanishing of the sum (23) and do not give the necessary extension of the definition of the components ω entering into (24).

PROPOSITION 5. *The sum* (23) *vanishes under the conditions*

$$u_{0,s} = u_{\tau N,s} = 0, \qquad v_{k,0} = v_{k,\tau M} = 0. \ \blacksquare \tag{26}$$

Conditions (26) look like (25), but they do not ensure the necessary extension of the definition of ω either and are beyond the framework of the scheme being used. Below we shall manage to pass to conditions (26) in the special case $g = 0$ by employing the orthogonal decompositions of the space H^1 (cf. 2.2).

The transition to conditions (26) is possible in the general case by the use of the discrete analog of the constructions carried out in 4.1, Ch. 2, namely, the glueing of the double and even-and-odd extensions of the forms.

3.4. Orthogonal decompositions. Let us now use one of the variants of the reasoning presented in 2.2. We denote by $\mathcal{D}_d^0$ (by $\mathcal{D}_\delta^0$) the linear space of the 0-forms (or 2-forms) that satisfy conditions

(15) (with the replacement of φ by η, respectively) and by $\mathfrak{N}_d^0$ (by $\mathfrak{N}_\delta^0$) the subspaces of H^1 consisting of elements representable in the form

$$\omega = d\varphi, \quad \varphi \in \mathcal{D}_d^0, \qquad \omega = \delta\eta, \quad \eta \in \mathcal{D}_\delta^0.$$

Then we can write for H^1 the decompositions

$$H^1 = \mathfrak{N}_d^0 \oplus \mathfrak{N}_\delta, \qquad H^1 = \mathfrak{N}_\delta^0 \oplus \mathfrak{N}_d, \tag{27}$$

where $\mathfrak{N}_\delta$, $\mathfrak{N}_d$ denote the orthogonal complements of the corresponding subspaces.

PROPOSITION 6. *The inclusions*

$$\mathfrak{N}_d^0 \subset \mathfrak{N}_\delta, \qquad \mathfrak{N}_d^0 \subset \mathfrak{N}_d$$

are valid.

PROOF. Let us verify the first inclusion. The condition $\omega \in \mathfrak{N}_d^0$ means that the components u, v can be represented as

$$u_{k,s} = \Delta_k \varphi_{k,s}, \qquad v_{k,s} = \Delta_s \varphi_{k,s},$$

where $\varphi \in \mathcal{D}_d^0$. Consequently, if the element $\omega \in H^1$ is orthogonal to $\mathfrak{N}_d^0$, then the relation

$$\sum_{k,s} [u_{k,s}(\Delta_k \varphi_{k,s}) + v_{k,s}(\Delta_s \varphi_{k,s})] = 0$$

is satisfied for any $\varphi \in \mathcal{D}_d^0$. This relation can be reduced to the form

$$-[\varphi_{k,s}(u_{k,s} + v_{k,s})]|_{\substack{k=1;N \\ s=1;M}} -$$

$$\sum_{s=2}^{M-1} [\varphi_{1,s}(\Delta_s v_{1,\sigma s} + u_{1,s}) + \varphi_{N,s}(\Delta_s v_{N,\sigma s} + u_{N,s})] -$$

$$\sum_{k=2}^{N-1} [\varphi_{k,1}(\Delta_k u_{\sigma k,1} + v_{k,1}) + \varphi_{k,M}(\Delta_k u_{\sigma k,M} + v_{k,M})] -$$

$$\sum_{k=2}^{N-1} \sum_{s=2}^{M-1} \varphi_{k,s}(\Delta_k u_{\sigma k,s} + \Delta_s v_{k,\sigma s}) = 0, \tag{28}$$

where now every factor $\varphi_{k,s}$ is encountered only once and assumes arbitrary values. Consequently, $\omega \in \mathfrak{N}_\delta$ if and only if every parenthesis entering into (28) vanishes. It is easy to verify that this condition is satisfied for $\omega \in \mathfrak{N}_\delta^0$.

The second inclusion can be verified similarly. ■

Decompositions (27) and Proposition 6 imply the following theorem.

THEOREM 3. *The space H^1 can be represented as the orthogonal sum*

$$H^1 = \mathfrak{R}_d^0 \oplus \mathfrak{R}_\delta^0 \oplus \mathfrak{N}_\Delta, \qquad \mathfrak{N}_\Delta = \mathfrak{N}_d \cap \mathfrak{N}_\delta. \blacksquare \tag{29}$$

As would be expected, it follows from our considerations that the relations

$$\delta\vartheta | x_{k,s} = 0, \qquad d\vartheta | V_{k,s} = 0,$$

are satisfied for the 1-forms $\vartheta \in \mathfrak{N}_\Delta$ on the elements $x_{k,s}$, $V_{k,s}$, lying inside V, i.e., ϑ is "harmonic inside V".

Let us now recall Eqs. (3) and their notation:

$$d\omega = f, \qquad \delta\omega = 0. \tag{3}$$

DEFINITION. We say that the *solution of Eqs.* (3) *under conditions* (25) is the 1-form $\omega \in \mathfrak{R}_d^0$ defined by the relations

$$\omega = \delta\eta, \qquad d\delta\eta = f, \qquad \eta \in \mathcal{D}_\delta^0, \tag{30}$$

which specify η as the solution of the Dirichlet problem in V. When calculating $\delta\eta$, we again take into account conditions (15).

The last stipulation may seem to be superfluous, but, strictly speaking, the solution of the Dirichlet problem defines the element H^2 and the requirement $\eta \in \mathcal{D}_\delta^0$ contains the restrictions imposed on the components η in the "neighborhood" of V. A similar situation may be encountered in connection with the fulfillment of conditions (25). The definition gives $\omega \in H^1$, but the first of equations (30) makes it possible to verify, on the assumptions made, the validity of relations (25).

The definition we have introduced and the preceding reasoning imply the following theorem.

THEOREM 4. *The solution of problem* (3), (25), *defined in the above-indicated way, exists for any $f \in H^2$ and is unique.*

For a complete justification of the definition that we have used, we should have verified that every 1-form ω, whose components satisfy relations (24) for $g = 0$ and conditions (25), is a solution in the sense having been described. This is actually true. ■

3.5. Step functions. In order to establish the relationship between the above-indicated purely combinatorial considerations and continual objects, we shall realize the scheme similar to that given in Sec. 1, Ch. 0.

Suppose that the domain $\Omega \subset \mathbb{R}^2$ is a rectangle with vertices (a_1, b_1), (a_2, b_1), (a_1, b_2), (a_2, b_2) defined by their Cartesian coordinates $0 \leq a_1 < a_2$, $0 \leq b_1 < b_2$. We introduce a *scale* h setting $h = N^{-1}(a_2 - a_1) = M^{-1}(b_2 - b_1)$. We shall consider the partitioning of the plane by the straight lines

$$x = a_1 + ph, \quad y = b_1 + qh, \qquad p, q = 0, \pm 1, \pm 2, \ldots, \tag{31}$$

and denote by $x_{k,s}$ the point of intersection of lines (31) for $p = k$, $q = s$. We denote by $V_{k,s}$ an open square bounded by the lines $p = k, \tau k$, $q = s, \tau s$. We denote the intervals of the straight lines (31) with the endpoints $x_{k,s}$, $x_{\tau k,s}$, $x_{k,s}$, $x_{k,\tau s}$ by $e^1_{k,s}$, $e^2_{k,s}$ and identify the geometric objects we have introduced with the combinatorial objects that we have considered and identify Ω with V. Note that in this way we fix the **orientation** of $\mathbb{R}^2$.

We associate every form $\varphi \in H^0$, $\omega \in H^1$, $\eta \in H^2$, defined on the corresponding discrete structure (on the complex $K(2)$ which is a conjugate of $\mathfrak{C}(2)$), with the step function (the pair of functions for $\omega = (u, v)$) assuming that

$$\varphi^h(x, y) = \varphi_{k,s}, \qquad u^h(x, y) = u_{k,s},$$

$$v^h(x, y) = v_{k,s}, \qquad \eta^h(x, y) = \eta_{k,s},$$

for $x, y \in V_{k,s}$. We shall regard the step functions defined in this way in Ω (or in V) as elements of the Hilbert spaces $\mathbb{H}^p$, $p = 0, 1, 2$, of square-summable functions (forms). It is obvious that

$$|\varphi^h, \mathbb{H}^0| = h|\varphi, H^0|$$

and similar relationships between norms are valid for ω^h, η^h. We define difference operators acting on the functions we have introduced, assuming that

$$\Delta^h_x \varphi^h(x, y) = h^{-1}[\varphi^h(x + h, y) - \varphi^h(x, y)],$$

$$\Delta^h_y \varphi^h(x, y) = h^{-1}[\varphi^h(x, y + h) - \varphi^h(x, y)],$$

and similarly define them for u^h, v^h, η^h. It stands to reason that these operators are defined in the **open** squares $V_{k,s}$. In the sequel it is always implied, although is not explicitly stipulated.

Let us define the action of the difference operators d^h, δ^h on the step forms φ^h, ω^h, η^h, replacing the partial derivatives D_x, D_y appearing in d, δ in a continual case (cf. notation (1) of Eqs. (3)) by the corresponding difference operators Δ_x^h, Δ_y^h.

PROPOSITION 7. *If relations of the form*

$$h\omega = d\varphi, \qquad h\eta = d\omega, \qquad h\omega = \delta\eta, \qquad h\varphi = \delta\omega$$

are satisfied for the discrete forms defined over the structure having been described, then the respective relations that contain the operators d^h, δ^h *are satisfied pointwise (this time without the factor* h*) for the step functions* φ^h, ω^h, η^h *constructed in accordance with the indicated rule.*

PROOF. Let, for instance, $h\omega = d\varphi$. This means that the relation $hu^h(x,y) = hu_{k,s} = \varphi_{\tau k,s} - \varphi_{k,s}$ is satisfied for $x, y \in V_{k,s}$. Then we simultaneously have $u^h(x,y) = h^{-1}[\varphi^h(x+h,y) - \varphi^h(x,y)]$ for $x, y \in V_{k,s}$.

The verification in the other cases is similar. ■

Let us now define the converse process, i.e., the association of a discrete object with a continual one. Suppose that $f(x,y)$ is a scalar function defined over a certain collection of squares $V_{k,s}$. We associate f with the step function f^h setting

$$f^h(x,y) = h^{-2} \int\limits_{V_{k,s}} f(x,y)\,dxdy \quad \text{for} \quad x, y \in V_{k,s}. \tag{32}$$

If f defined a 0-form, namely, the element $\mathbb{H}^0$, a component of the element $\mathbb{H}^1$, or the element $\mathbb{H}^2$, then we can associate it with the corresponding discrete object assigning the value f^h to the point $x_{k,s}$, to one of the intervals $e^1_{k,s}$, $e^2_{k,s}$, or to the element $V_{k,s}$. We call this procedure *discretization.*

Proposition 7 and law (32) allow us to formulate and verify Theorem 5 given below. Results of this kind serve as the basis of the construction of the approximations of the solutions of the problems which are of interest to us.

We introduce the notation

$$|\varphi, \mathcal{W}|^2 = \sum_{k=0}^{N}\sum_{s=0}^{M}[(\Delta_k\varphi_{k,s})^2 + (\Delta_s\varphi_{k,s})^2] = |\varphi^h, W|^2, \tag{33}$$

where a discrete form is on the left-hand side and a step function is on the right-hand side.

In order to verify whether the norms $\mathcal{W}$, W, introduced in this way, are natural, it is sufficient to multiply and divide the middle term by h^2 (here we make use of the specific properties of a two-dimensional case). Since the summation begins with $k = 0$, $s = 0$, we have taken into account the additional terms appearing in (17).

THEOREM 5. *Suppose that the step function g^h is the discretization of $g \in \mathbb{H}^0$. Then the Dirichlet problem for the equation*

$$\delta^h d^h \varphi^h = g^h \tag{34}$$

in V is uniquely solvable and the inequality

$$|\varphi^h, W| < c|g, \mathbb{H}^0| \tag{35}$$

is satisfied for the solution φ^h.

PROOF. We understand the solution of the Dirichlet problem for (34) as the step function φ^h constructed in accordance with the discrete form φ, i.e., with the solution of the corresponding problem for the equation

$$\delta\, d\varphi = h^2\hat{g},$$

where $\hat{g}$ is a 0-form defined by the step function g^h and the factor h^2 corresponds to the transition from d^h, δ^h to d, δ.

The unique solvability follows from Theorem 1. Setting $\psi = \varphi$, we find from (14), (17) that

$$|\varphi, \mathcal{W}|^2 < h^2\left|\sum_{k,s}\hat{g}_{k,s}\varphi_{k,s}\right| \le |g^h, \mathbb{H}^0|\,|\varphi^h, \mathbb{H}^0|, \tag{36}$$

but

$$\varphi_{k,s}^2 = \left(\sum_{p=0}^{k}\Delta_p\varphi_{p,s}\right)^2 \le k\sum_{p=0}^{N}(\Delta_p\varphi_{p,s})^2$$

and summing up over k, we get

$$\sum_{k}\varphi_{k,s}^2 \le N^2\sum_{p=0}^{N}(\Delta_p\varphi_{p,s})^2.$$

Summing up over s and dividing by $N^2 = h^{-2}$, we obtain

$$|\varphi, \mathbb{H}^0|^2 \leq \sum_s \sum_{k=0}^{N} (\Delta_k \varphi_{k,s})^2 \leq |\varphi^h, W|^2. \tag{37}$$

It is easy to obtain the estimate

$$|g^h, \mathbb{H}^0| \leq c|g, \mathbb{H}^0|. \tag{38}$$

Relations (33), (36), (37), (38) imply (35). ■

Similar statements that can be verified in the same way are valid for the equations (18), (19). In the last case we must naturally use (22).

3.6. Approximation and the limiting process. Proceeding now from step function, we shall construct "smooth" (of the class C^1 or C^2) functions which give, in a certain sense, the approximation of solutions of fundamental equations. For the beginning, we concentrate our attention on the case of the step 0-form (a scalar function) φ^h. Suppose that

$$J^h \varphi^h(x,y) = h^{-2} \int_x^{x+h} \int_y^{y+h} \varphi^h(\xi, \eta)\, d\xi d\eta, \qquad h > 0,$$

is Steklov's function with the averaging radius equal to the parameter h of the net. The latter condition is not necessary for the realization of the given constructions, but is convenient. Note that the function $J^h\varphi^h$ is an example of the simplest spline.

PROPOSITION 8. *If $\varphi_{k,s} \in H^0$ satisfies condition* (15), *then the function $J^h\varphi^h(x,y)$ belongs to the space $\dot{W}^1$ in $\Omega_h = (a_1 - h, a_2) \times (b_1 - h, b_2)$.*

The statement is obvious since $J^h\varphi^h$ is absolutely continuous with respect to the totality of the variables, is piecewise differentiable, and satisfies homogeneous boundary conditions. ■

Let us now consider the infinite increasing sequence N_q, M_q, $q = 0, 1, \ldots$: $N_0 = N$, $M_0 = M$, $N_1 = 2N$, $M_1 = 2M$, ..., and the corresponding sequence $h_q \to 0$ supposing that the constructions described above have been realized for every one of the values h_q. In what follows, we shall not explicitly indicate the dependence of

h on q, but, speaking of a certain "collection of values of h" or of a "family of functions φ^h", we shall imply the indicated sequence.

PROPOSITION 9. *Suppose that g is a fixed element of $\mathbb{H}^0$ and $\{\varphi^h\}$ is a family of solutions of the Dirichlet problem for Eqs.* (34) *which corresponds to different values of h. Then the norms in $\mathbb{H}$ of the generalized derivatives $D_x J^h \varphi^h$, $D_y J^h \varphi^h$ are bounded uniformly with respect to h.*

PROOF. Let us consider

$$D_x J^h \varphi^h = h^{-1} \int_y^{y+h} \Delta_x^h \varphi(x, \eta)\, d\eta = J_y^h \Delta_x^h \varphi^h(x, y),$$

where J_y^h means the averaging with respect to one variable y. The smoothing operator which acts on all as well as on a part of the variables, is always bounded uniformly with respect to h, i.e.,

$$|J_y^h \Delta_x^h \varphi^h(x, y)\mathbb{H}^0| \leq c|\Delta_x^h \varphi^h, \mathbb{H}^0|,$$

and our statement follows from Theorem 5 (when we take the differences, we can consider $J^h \varphi^h$ to be continued by zero for $x > a_2$, $y > b_2$).

PROPOSITION 10. *The family $\{J^h \varphi^h\}$, where $\{\varphi^h\}$ are solutions of the Dirichlet problem for the fixed right-hand side considered in Proposition* 9, *is compact in $\mathbb{H}^0(\Omega)$.*

This statement follows from Proposition 9 and the compactness of the inclusion $\dot{W}^1 \subset \mathbb{H}$. Although $J^h \varphi^h$ are zero, strictly speaking, only on a part of $x = a_2$, $y = b_2$ of the boundary Ω, this is sufficient for the statement to be valid. ■

PROPOSITION 11. *The estimate*

$$|J^h \varphi^h - \varphi^h, \mathbb{H}^0(\Omega)| \leq c|h|,$$

where the constant c is independent of h, is valid for the family $\{\varphi^h\}$ of solutions of the Dirichlet problem from Proposition 9.

PROOF. Let $x \in V_{k,s}$. Then, calculating the corresponding integral, we have

$$J^h \varphi^h = h^{-2}\{\varphi_{k,s}|V^1| + \varphi_{\tau k,s}|V^2| + \varphi_{k,\tau s}|V^3| + \varphi_{\tau k,\tau s}|V^4|\},$$

where $|V^\mu|$ are the areas of the parts of the corresponding squares ($V_{k,s}$ and the adjacent ones), that belong to the integration domain, and $h^{-2}\sum_1^4 |V^\mu| = 1$. Using the representation

$$\varphi_{k,s} = h^{-2}\sum_1^4 \varphi_{k,s}|V^\mu|$$

and noting that $h^{-2}|V^\mu| \le 1$, we obtain

$$|J^h\varphi^h - \varphi^h|_{V_{h,s}} \le |\varphi_{\tau k,s} - \varphi_{k,s}| + |\varphi_{k,\tau s} - \varphi_{k,s}| + |\varphi_{\tau k,\tau s} - \varphi_{k,s}|,$$

whence it follows that

$$\int_\Omega |J^h\varphi^h - \varphi^h|^2\,dx\,dy \le ch^2\sum_{h,s}\{(\varphi_{\tau k,s} - \varphi_{k,s})^2 + (\varphi_{k,\tau s} - \varphi_{k,s})^2\}.$$

Since, on our assumptions, the sum on the right-hand side is bounded (relation (33) and Theorem 5), we get what we required. ■

PROPOSITION 12. *The family $\{\varphi^h\}$ of solutions of the Dirichlet problem from Proposition* 9 *is compact in* $\mathbb{H}^0(\Omega)$.

PROOF. We fix an arbitrary $\varepsilon > 0$ and prove the existence of a finite ε-net for the family $\{\varphi^h\}$. As was proved, the family $\{J^h\varphi^h\}$ is compact in $\mathbb{H}^0(\Omega)$ and, consequently, there exists a finite $\varepsilon/2$-net $\{\nu_k\}$ for it. At the same time, in accordance with Proposition 11, there exists h_q such that

$$|J^h\varphi^h - \varphi^h, \mathbb{H}^0| < \varepsilon/2 \quad \text{for} \quad h \le h_q,$$

so that $\{\nu_k\}$ is a finite ε-net for $\{\varphi^h\}_{h\le h_q}$. Adding the elements for $\{\varphi^h\}_{h\ge h_q}$, whose number is finite, to the net $\{\nu_k\}$, we obtain a finite ε-net for the whole family $\{\varphi^h\}$. ■

In what follows, when speaking of the sequence of solutions of the Dirichlet problem (with the fixed right-hand side), we shall mean the existing (according to what was proved) sequence which converges in $\mathbb{H}^0$ to a certain element $\varphi \in \mathbb{H}^0$.

REMARK. It stands to reason that in actual fact the whole sequence $\{\varphi^{h_q}\}$ converges to φ. However, it turns out to be difficult to carry out a direct accurate verification of this fact. We shall establish this fact later.

PROPOSITION 13. *The sequence $\{J^h\varphi^h\}$ strongly converges to φ.*

It is sufficient to note that

$$|J^h\varphi^h - \varphi, \mathbb{H}^0| \le |J^h\varphi^h - \varphi^h, \mathbb{H}^0| + |\varphi^h - \varphi, \mathbb{H}^0|. \blacksquare$$

PROPOSITION 14. *The element φ belongs to the space $\dot{W}^1$.*

PROOF. Proposition 9 is equivalent to the well-known criterion of existence of the so-called **weak** generalized derivatives $D_x\varphi$, $D_y\varphi$. Recall the corresponding reasoning [43]. According to Proposition 9, there exists a sequence for $\{D_x J^h \varphi^h\}$ which weakly converges to a certain $\chi \in \mathbb{H}$. In addition, for the function $\psi \in C^1$ we have

$$\int D_x\psi \cdot J^h\varphi^h \, dx\, dy = -\int \psi D_x J^h \varphi^h \, dx\, dy,$$

where, taking into account our agreement concerning the continuation of $J^h\varphi^h$ by zero, we can consider the integration to be extended to the whole $\mathbb{R}^2$. Passing to the limit, we obtain

$$\int D_x\psi \cdot \varphi \, dx\, dy = -\int \psi\chi \, dx\, dy,$$

i.e., $D_x^{\mathrm{wk}}\varphi = \chi$, but in the domain under consideration the weak and strong definitions of differentiability (and the fulfillment of the boundary conditions) are equivalent (cf. [19]). The verification of the existence of $D_y\varphi$ is similar. ■

PROPOSITION 15. *For any $\psi \in C^2$, $\psi|_{\partial\Omega} = 0$ the relation*

$$(\varphi, \delta d\, \psi) = (g, \psi) \tag{39}$$

is satisfied for the element φ we have found.

PROOF. Using Theorem 5, we can write

$$(g^h, \psi) = (\delta^h d^h \varphi^h, \psi) = (\varphi^h, \delta^h d^h \psi).$$

Passing to the limit as $h \to 0$, we obtain (39). ■

PROPOSITION 16. *The obtained element φ is a $\dot{W}^1$-solution of the Dirichlet problem, i.e., the relation*

$$(d\varphi, d\psi) = (g, \psi)$$

is satisfied for any $\psi \in \dot{W}^1$.

PROOF. Writing (39) for the W^2-approximation ψ_ε of the element ψ,

$$|\psi_\varepsilon - \psi, \dot{W}^1| \to 0 \quad \text{as} \quad \varepsilon \to 0,$$

transferring the operator δ to $\varphi \in \dot{W}^1$, and passing to the limit as $\varepsilon \to 0$, we get what we required. ■

PROPOSITION 17. *The $\dot{W}^1$-solution of the Dirichlet problem is unique.*

In this situation this standard result can be regarded as a consequence of relation (17). ■

COROLLARY. *The whole sequence $\{\varphi^{h_q}\}$ strongly converges to φ.*

Indeed, if we assumed the contrary, we would construct one more $\dot{W}^1$-solution of the Dirichlet problem different from φ. ■

The following theorem is a final result of our considerations.

THEOREM 6. *The sequence $\{\varphi^h\}$ of step functions, constructed on the basis of the given element $g \in \mathbb{H}^0$, strongly converges in $\mathbb{H}^0$ to the element $\varphi \in \dot{W}^1(\Omega)$, i.e., to the generalized solution of the problem*

$$-\Delta\varphi = g, \qquad \varphi\Big|_{\partial\Omega} = 0. \tag{40}$$

At the same time, the sequence $\{J^h\varphi^h\}$ converges to φ in the metric $\dot{W}^1$. ■

Let us turn to system (1) written in the form (3):

$$d\omega = f, \qquad -\delta\omega = 0. \tag{3}$$

In order to use the scheme of constructing the solution of this system, similar to that given in 3.4, we need the following statement.

PROPOSITION 18. *In the statement of Theorem 6 we can replace the sequence $\{J^h\varphi^h\}$ by $\{(J^h)^2\varphi^h\}$, where $(J^h)^2$ is the iteration of J^h.* ■

REMARK. It should be emphasized that although $(J^h)^2\varphi^h \in C^2$, we by no means assert that $(J^h)^2\varphi^h$ converges to φ with respect to the W^2-norm, i.e., that φ is the W^2-solution of problem (40).

The validity of the statement is a simple consequence of the averaging properties.

It should be mentioned in passing that the function $(J^h)^2\varphi^h$ is a parabolic spline.

PROPOSITION 19. *As applied to the Dirichlet problem for the equation*

$$d\delta\eta = f \in \mathbb{H}^2 \tag{41}$$

the analogs of Theorem 6 and Proposition 18 hold true.

We have repeatedly mentioned the validity of this transition from φ to η and vice versa. ■

DEFINITION. We call the element $\omega \in \mathbb{H}^1(\Omega)$ a *generalized solution of problem* (1)–(2) in Ω if

$$(\omega, \delta\psi) = (f, \psi) \tag{42}$$

for any $\psi \in \dot{W}^1(\Omega)$ and, at the same time, there exists a sequence $\omega(h) \in W^1$ such that $\omega(h) \to \omega$ in $\mathbb{H}^1$ as $h \to 0$ and the relation

$$\delta\omega(h) = 0, \tag{43}$$

is satisfied for $h > 0$, $\omega(h)$ satisfying the boundary conditions (2).

THEOREM 7. *The generalized solution of problem* (1)–(2), *defined in the above-indicated way, exists and is unique for any* $f \in \mathbb{H}^2$.

UNIQUENESS. By virtue of our assumptions, the equation $\delta\psi(h) = \omega(h)$ is solvable in $\dot{W}^1$ for $h > 0$. Setting $f = 0$, $\psi = \psi(h)$ in (42), we find that $(\omega, \omega(h)) = 0$. Passing to the limit as $h \to 0$, we get what we required.

EXISTENCE. We set $\omega(h) = \delta(J^h)^2\eta^h$, where $\{\eta^h\}$ is a family of solutions of the Dirichlet problem for the step analog of Eq. (41) (we use the analog of Eq. (34) for (41) and Proposition 19). For any $\psi \in \dot{W}^1$ we have

$$(d\omega(h), \psi) = (\omega(h), \delta\psi) = (\delta(J^h)^2\eta^h, \delta\psi).$$

However, in these relations we can pass to the limit as $h \to 0$, and this gives (42). In this case relation (43) and the boundary conditions are identically satisfied. ∎

4. Discrete Analogs of Certain Hydrodynamics Relations

4.0. Preliminary remarks. It is most convenient to elucidate the essence of this section by using analogies. The Poisson difference equation

$$-\Delta\varphi = f \tag{1}$$

for the function (scalar) φ, written over the unique point $x_{0,0}$ of our model of $\mathbb{R}^2$, has the form

$$4\varphi_{0,0} - \varphi_{1,0} - \varphi_{0,1} - \varphi_{-1,0} - \varphi_{0,-1} = f_{0,0} \tag{2}$$

(as in the first part of Sec. 3, we are not interested in the "scale" of h and the corresponding normalization of the differences). The unique solvability of the "Dirichlet problem" for (2),

$$\varphi_{1,0} = \varphi_{0,1} = \varphi_{-1,0} = \varphi_{0,-1} = 0, \tag{3}$$

is obvious. Moreover, setting $f = 0$, we have the "mean value theorem"

$$\varphi_{0,0} = \frac{1}{4}(\varphi_{1,0} + \varphi_{0,1} + \varphi_{-1,0} + \varphi_{0,-1})$$

and the "maximum principle"

$$|\varphi_{0,0}| \leq \max_{k,j} |\varphi_{k,j}|$$

for the "harmonic" φ.

Now if we want to simulate in this way some properties of the Neumann problem for (1), then we shall have to add to (2) the equation at the adjacent point,

$$4\varphi_{1,0} - \varphi_{0,0} - \varphi_{1,1} - \varphi_{2,0} - \varphi_{1,-1} = f_{1,0}, \tag{4}$$

writing the conditions

$$\varphi_{-1,0} - \varphi_{0,1} = \varphi_{0,1} - \varphi_{0,0} = \varphi_{1,1} - \varphi_{1,0} =$$

$$\varphi_{2,0} - \varphi_{1,0} = \varphi_{1,-1} - \varphi_{1,0} = \varphi_{0,-1} - \varphi_{0,0} = 0.$$

instead of (3). The resulting pair of equations $\varphi_{0,0} - \varphi_{1,0} = f_{0,0} = -f_{1,0}$ is solvable only when $f_{0,0} + f_{1,0} = 0$ (the "orthogonality of f to the constants") and defines φ with an accuracy to within the constant.

In this section the simplest "natural" patterns (similar to the "5-point" and "8-point" patterns for (1)) are indicated for the discrete analog of the equations

$$d * \mathfrak{U} = 0, \qquad \mathfrak{U} \wedge * d\mathfrak{U} = * d\gamma \tag{5}$$

(cf. (2), (6), Sec. 6, Ch. 2) and difference analogs of some relations from 6.1, Ch. 2 are written. Of decisive importance is the use of the $\smile$-multiplication instead of the exterior one which makes it possible to "correctly" simulate quasilinearity.

Another variant, less obvious, based on the notation of the second equation of (5) in the form

$$(J\mathfrak{U}) \lrcorner \, d\mathfrak{U} + d\gamma = 0$$

(see the end of 6.1, Ch. 2) was considered in [16].

No attempt is made in this book to investigate the constructed analog for (5) in a "large" domain and carry out a limiting process.

4.1. The minimal natural pattern. Using the constructed discrete model of the Euclidean space, we shall consider the simplest pattern and the difference relations for our Eq. (5), which are analogs of the 5-point pattern for Eq. (1) and in some respects of Eqs. (2), (3) and the obtained characteristics of the corresponding solution.

We shall use the notations and agreements introduced in 3.2 for the model of $\mathbb{R}^2$, changing slightly the notations of concrete objects (bringing them into correspondence with the notations of 6.1, Ch. 2). The vector $\mathfrak{U} = (u_{k,s}, v_{k,s})$ will be the stream vector. Proceeding from the element of the volume $V_{1,1}$, we have

$$\begin{aligned} d * \mathfrak{U} \,|\, V_{1,1} &= u_{1,1} - u_{0,1} + v_{1,1} - v_{1,0}, \\ d\mathfrak{U} \,|\, V_{1,1} &= v_{2,1} - v_{1,1} - u_{1,2} + u_{1,1}. \end{aligned} \tag{6}$$

Taking into account that the value of the curl $\omega = *d\mathfrak{U}$ appearing in the equations coincides with the value $d\mathfrak{U}$ (but corresponds to the point $x_{2,2}$),

$$\omega_{2,2} = v_{2,1} - v_{1,1} - u_{1,2} + u_{1,1},$$

using the rules of the exterior multiplication of a covector by a scalar, and introducing into consideration the scalar γ that denotes the pressure, we obtain

$$u_{1,2}\,\omega_{2,2} = \gamma_{1,1} - \gamma_{1,2}, \qquad v_{2,1}\,\omega_{2,2} = \gamma_{2,1} - \gamma_{1,1} \tag{7}$$

(we have written the values of $\mathfrak{U} \wedge \omega$ and $* \, d\gamma$ on $e^1_{1,2}$ and $e^2_{2,1}$) and

$$u_{1,1} - u_{0,1} + v_{1,1} - v_{1,0} = 0, \tag{8}$$

i.e., three equations for nine quantities, the functions on the elements of the corresponding dimension in our model (see the figure).

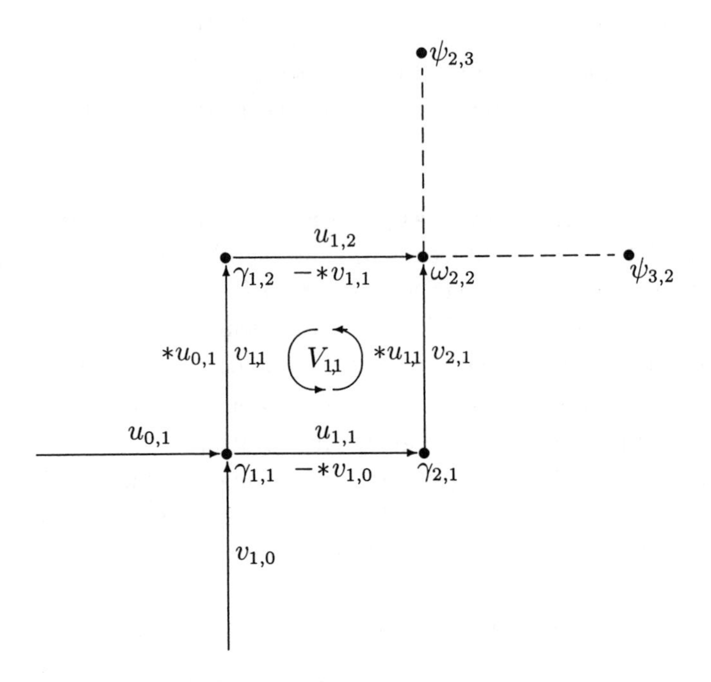

Being "minimal", the system of Eqs. (7), (8) is very poor and does not reflect, in particular, some specific properties of solvability of Eqs. (5), which we shall consider in the next subsection having "tripled" the pattern. Nevertheless, some remarks are due here.

Clearly, we can always obtain the simplest class of "flows" (potential flows) by setting $\omega = 0$. Equating relations (6) to zero, we obtain the difference analog of the Cauchy–Riemann equations for "covector functions".

Eqs. (7) immediately yield the relation

$$(\gamma_{2,1} - \gamma_{1,1})u_{1,2} + (\gamma_{1,2} - \gamma_{1,1})v_{2,1} = 0,$$

corresponding to the Bernoulli law.

Let us now discuss the introduction of the scalar ψ, the stream function: $*\mathfrak{U} = d\psi$. We shall have to include the points $x_{2,3}$, $x_{3,2}$ into the number of geometric elements of the pattern. We get

$$u_{0,1} = \psi_{1,2} - \psi_{1,1}, \qquad u_{1,1} = \psi_{2,2} - \psi_{2,1}$$
$$u_{1,2} = \psi_{3,2} - \psi_{2,2} \tag{9}$$

and the corresponding relations which define $v_{1,0}$, $v_{1,1}$, $v_{2,1}$. We can now regard Eqs. (7), (8) as those for the function γ and the new unknown function ψ. Eq. (8) will be identically satisfied and Eqs. (7) become nonlinear relations for ψ containing the factor

$$-\Delta\psi \,|\, x_{2,2} = 4\psi_{2,2} - \psi_{3,2} - \psi_{2,3} - \psi_{2,1} - \psi_{1,2} = \omega_{2,2}. \tag{10}$$

The "poverty" of the minimal pattern tells on the possibility of specifying (when considering the equations for ψ) the arbitrary value $\omega_{2,2}$. At the same time, the specific feature of Euler's equations is the existence of additional nonlinear restrictions for ω the relation

$$\omega = \Delta\psi \tag{11}$$

(in contrast to 6.1, Ch. 2, it is convenient now to change the sign of ψ in (11)). In 4.2 we shall obtain the analog for these restrictions.

4.2. A tripled pattern and solvability. Let us now "triple" the pattern that we have introduced reproducing the constructions of the preceding subsection in the case of three volume elements $V_{1,1}$, $V_{1,2}$, $V_{2,1}$. In addition to the three equations for divergence which are analogs of (8),

$$d * \mathfrak{U} \,|\, V_{1,1} = d * \mathfrak{U} \,|\, V_{1,2} = d * \mathfrak{U} \,|\, V_{2,1} = 0, \tag{12}$$

we get a chain of six equations, which are analogs of the pair (7):

$$\begin{gathered} u_{1,2}\omega_{2,2} = \gamma_{1,1} - \gamma_{1,2}, \qquad u_{1,3}\omega_{2,3} = \gamma_{1,2} - \gamma_{1,3}, \\ u_{2,2}\omega_{3,2} = \gamma_{2,1} - \gamma_{2,2}, \\ v_{2,1}\omega_{2,2} = \gamma_{2,1} - \gamma_{1,1}, \qquad v_{2,2}\omega_{2,3} = \gamma_{2,2} - \gamma_{1,2}, \\ v_{3,1}\omega_{3,2} = \gamma_{3,1} - \gamma_{2,1}. \end{gathered} \tag{13}$$

Now these equations are no longer independent, i.e., excluding $\gamma_{1,1}$ from the first two equations and $\gamma_{2,2}$ from the fourth and fifth equations, we obtain the basic nonlinear relation

$$(v_{2,1} + u_{1,2})\omega_{2,2} = v_{2,2}\omega_{2,3} + u_{2,2}\omega_{3,2}. \tag{14}$$

When we exclude one of the equations, that we have used, from the chain (13), the remaining five equations define (for the specified u, v, ω) the values of γ with an accuracy to within the constant term. As to Eq. (14), it must be used, in combination with the three equations (12), for defining $\mathfrak{U}$.

It is more convenient to investigate the situation that occurs by using the stream function ψ. After adding the points $x_{2,4}$, $x_{3,3}$, $x_{4,2}$ to the elements of the pattern, we must add

$$u_{0,2} = \psi_{1,3} - \psi_{1,2}, \quad u_{1,3} = \psi_{2,4} - \psi_{2,3},$$

$$u_{2,1} = \psi_{3,2} - \psi_{3,1}, \quad u_{2,2} = \psi_{3,3} - \psi_{3,2}$$

to relations (9) for u and write out the corresponding relations for v. Eqs. (12) for the new unknown function ψ will be identically satisfied. In order to use the analog of relation (11), we add relations

$$\Delta\psi \,|\, x_{2,3} = \omega_{2,3}, \qquad \Delta\psi \,|\, x_{3,2} = \omega_{3,2} \tag{15}$$

to (10) and construct a "flow" by adding the "Dirichlet conditions" for ψ,

$$\psi_{1,1} = \psi_{2,1} = \psi_{3,1} = \psi_{1,2} = \psi_{4,2} = \psi_{1,3} = \psi_{3,3} = \psi_{2,1} = 0,$$

to (10), (15). Now we can define $\mathfrak{U}$ by specifying in (10), (15) either the values of ω or the values of the three quantities $\psi_{2,2}$, $\psi_{3,2}$, $\psi_{2,3}$ that remain free. At the same time, the nonlinear compatibility conditions, following from (14), must be satisfied. We can write them in the form

$$(\psi_{3,2} - \psi_{2,3})(\psi_{3,2} + \psi_{2,3} - \psi_{2,2}) = 0. \tag{16}$$

REMARK. We have obtained conditions (16) directly from (13). We can verify that the use of the discrete analog of the relation $d\psi \wedge d\omega = 0$ leads to the same result.

Relation (16) can obviously also be written as restrictions imposed on ω, or restrictions imposed on the corresponding values of u, v. In this event, the structure of the restrictions remains unchanged.

Let us consider the variants of definition of the "flow" by the choice of ψ.

1. Let us specify $\psi_{3,2}$, $\psi_{2,3}$. The value $\psi_{2,2}$ is uniquely defined by (16) if $\psi_{3,2} \neq \psi_{2,3}$; if $\psi_{3,2} = \psi_{2,3}$, then $\psi_{2,2}$ remains arbitrary (we have a continuum of solutions).

2. We specify $\psi_{3,2}$, $\psi_{2,2}$ (or $\psi_{2,3}$, $\psi_{2,2}$). If $\psi_{2,2} \neq 2\psi_{3,2}$ ($\psi_{2,2} \neq 2\psi_{2,3}$), we can choose the value $\psi_{2,3}$ (or $\psi_{3,2}$) in two ways.

A similar situation occurs if we directly solve Eqs. (12), (13) (with some additional "boundary" conditions imposed on the components of $\mathfrak{U}$). When the number of elements entering into the scheme is increased, the number of nonlinear relations arising from the chain of type (13) increases and the investigation becomes more complicated.

Note that when we consider a "long" scheme that includes the elements of volume $V_{k,s}$, $k = 1, \ldots, M$, $s = 1, 2, 3$, under the boundary conditions $v_{k,1} = v_{k,3} = 0$, the preservation of the values of ω is verified directly on the sequence of elements of $V_{k,2}$ (the "curl is preserved").

4.3. Nonstationary equations. We shall show now that the situation connected with the solvability undergoes a qualitative change when "time" is added. The latter fact means the addition of the dependence of $\mathfrak{U}$ and γ (ψ, respectively) on the continuous parameter t. The inclusion of this dependence must correspond to the consideration of Euler's nonstationary homogeneous equations resulting from the addition of the derivative $D_t u$ to the first of Eqs. (1), Sec. 6, Ch. 2, and the derivative $D_t v$ to the second equation.

For our purposes, it is sufficient to add the corresponding derivatives to those of Eqs. (13) which participated in finding conditions (14). According to our rules (in fact, we apply the operator $*^{-1}$ to the equation $\mathfrak{U} \wedge *du = *d\gamma$), we have

$$
\begin{aligned}
D_t u_{1,2} - v_{2,2}\omega_{2,3} &= \gamma_{1,2} - \gamma_{2,2}, \\
D_t v_{1,1} + u_{1,2}\omega_{2,2} &= \gamma_{1,1} - \gamma_{1,2}, \\
D_t v_{2,1} + u_{2,2}\omega_{3,2} &= \gamma_{2,1} - \gamma_{2,2}, \\
D_t u_{1,1} - v_{2,1}\omega_{2,2} &= \gamma_{1,1} - \gamma_{2,1},
\end{aligned}
$$

whence it follows that

$$D_t v_{2,1} - D_t u_{1,2} + v_{2,2}\omega_{2,3} + u_{2,2}\omega_{3,2} = D_t v_{1,1} - D_t u_{1,1} + (u_{1,2} + v_{2,1})\omega_{2,2}.$$

When writing the analog of condition (16), it is convenient to introduce the notations

$$\psi_{2,3} - \psi_{3,2} = \alpha(t), \qquad \psi_{2,3} + \psi_{3,2} = \beta(t).$$

If we now consider α, β to be given functions of t (if we return to u, v, then this is, in a certain sense, the "specification of the curl at the input"), then $\psi_{2,2}$ can be found from the equation

$$4D_t\psi_{2,2} = 5\alpha(t)\psi_{2,2} - \alpha(t)\beta(t) + D_t\beta(t),$$

to which we must add the condition

$$\psi_{2,2}(0) = \psi_{2,2}^0$$

(the complete definition of the "flow" at the initial moment).

REMARK. Conditions (16) also essentially change their character when we include perturbations of the type of "viscosity." In order to make a reasonable account of the effect of the viscosity, it is necessary to introduce the normalization which distinguishes the "leading" derivatives.

5. A Double Complex

As was pointed out above, when modeling $\mathbb{R}^n$, we can introduce a combinatorial object, namely, a "double complex," in which the $*$ operation is defined in such a way that $** = 1$ (or $** = \pm 1$), i.e., it is more like its continual analog. Here are the corresponding constructions.

Together with the complex $\mathfrak{C}(n)$ (see 1.2) we shall consider its "double," namely, the complex $\tilde{\mathfrak{C}}(n)$ of exactly the same structure. We define the one-to-one correspondence

$$*: \ \mathfrak{C} \to \tilde{\mathfrak{C}}, \quad \tilde{\mathfrak{C}} \to \mathfrak{C} \tag{1}$$

(we write $\mathfrak{C}$ instead of $\mathfrak{C}(n)$), defining the corresponding map of the base elements:

$$*: \ s_{(p)} \mapsto \tilde{s}_{(n-p)}, \quad \tilde{s}_{(n-p)} \mapsto s_{(p)},$$

where the element $\tilde{s}_{(n-p)}$, which is an element of $\tilde{\mathfrak{C}}$, is transversal (by definition) for $s_{(p)} \in \mathfrak{C}$. Here is what we mean.

As in 1.2, we suppose that the subscript (p) carries all information about $s_{(p)}$, i.e.,

$$s_{(p)} = s_{\varkappa_1} \otimes \ldots \otimes s_{\varkappa_n}, \tag{2}$$

where $(\varkappa_1, \ldots, \varkappa_n)$ is the corresponding "point" and product (2) contains exactly p one-dimensional elements $e_{\varkappa_j}$ and $n-p$ zero-dimensional elements $x_{\varkappa_i}$. Then

$$* s_{(p)} = \tilde{s}_{(n-p)} = * \tilde{s}_{\varkappa_1} \otimes \ldots \otimes * \tilde{s}_{\varkappa_n},$$

and $* \tilde{s}_{\varkappa_j} = \tilde{e}_{\varkappa_j}$ if $s_{\varkappa_j} = x_{\varkappa_j}$ and $* \tilde{s}_{\varkappa_j} = \tilde{x}_{\varkappa_j}$ if $s_{\varkappa_j} = x_{\varkappa_j}$. The mapping $*: \ \tilde{\mathfrak{C}} \to \mathfrak{C}$ can be defined by analogy so that $** = 1$.

The complex of the cochains $\tilde{K}(n) = \tilde{K}$ over $\tilde{\mathfrak{C}}$, with the operator d defined in it (generated by the pairing and the operation ∂ in $\tilde{\mathfrak{C}}$) has the same structure as in $K(n)$, and, at the same time, the operation (1) induces the respective mapping

$$*: \ K \to \tilde{K}, \ \ \tilde{K} \to K$$

according to the law

$$\langle \tilde{a}, * \varphi \rangle = \langle * \, \tilde{a}, \varphi \rangle, \qquad \langle a, * \tilde{\psi} \rangle = \langle * \, a, \tilde{\psi} \rangle,$$

where $\varphi \in K$, $\tilde{\psi} \in K$.

As the basic domain we again take

$$V = \sum_{\varkappa} V_{\varkappa} \subset \mathfrak{C}(n),$$

where $V_{\varkappa}$ are n-dimensional base elements. However, in the definition of the inner product the part of the $\smile$-multiplication is now played by $\otimes$-multiplication (tensor multiplication). We set

$$V_p = \pm \sum_{(V)} s_{(p)} \otimes * \, s_{(p)},$$

where the sum is taken over all $s_{(p)}$ belonging to the p-dimensional skeleton of V. We can choose the sign (dependent on n and p) such that the orientation of the elements of V_p would coincide with the natural orientation of the elements of V (and this does not always automatically occur under our choice of the definition of $** = 1$). For u, $v \in K^p(n)$ we set

$$(u, v)_V = \langle V_p, u \otimes * v \rangle = \sum_{(V)} \langle s_{(p)}, u \rangle \langle * \, s_{(p)}, * \, v \rangle.$$

The inner product makes it possible to define the operator δ (the conjugate of d) and write the analog of the Green formula:

$$(du, v)_V = \langle \partial V, u \otimes * v \rangle + (u, \delta v)_V.$$

No essential modifications are needed to carry out constructions, given in Sec. 2, in a double complex [15]. The situation is more difficult when we deal with constructions from Sec. 3. Here the whole combinatorics must be recalculated anew. In particular, when duplicating 3.5, we have to deal with a "double net" of straight lines, i.e., in addition to those indicated in (31) we shall need straight lines defined by the relations

$$x = a_1 + \left(p + \frac{1}{2}\right)h, \quad y = b_1 + \left(q + \frac{1}{2}\right)h, \quad p, q = 0, \pm 1, \dots$$

Correspondingly, the limiting process also becomes more difficult. A "topological" construction of the type of the construction considered in 1.3, that suits this situation, was suggested in [15].

It is also not quite clear whether it is possible to carry out constructions from Sec. 4, which make a considerable use of the analog of exterior multiplication, in a double complex.

The merit of a double complex is the visualization which the operation $*$ acquires in it. This visualization sometimes proves to be useful when we develop numerical procedures (cf. [41]).

6. Some Open Problems

Here is a number of problems which are connected with the content of this chapter and which have not been cleared up. Some of them were mentioned above.

All the given constructions of a general nature (Sec. 2) mainly referred to the domain V which is a combinatorial analog of the domain (of a connected open set) $V \subset \mathbb{R}^n$ in $\mathfrak{C}(n)$. For instance, the question of generalizing our considerations to the case of a two-dimensional complex, namely, the boundary of a three-dimensional domain, remains open. In this case it is easy to determine "formal" inner products, but when we attempt to define the operation $*$, a number of problems arise. Possibly, it would be convenient to employ here a construction based on the use of a double complex. It is interesting to study problems of this type since, say, the question concerning the global difference approximations "throughout the surface of the earth" has not been studied throughly enough.

The choice of the domain V used in Sec. 3 is even more special. The extension of the results to "irregular" domains requires additional efforts (in particular, different versions of the generalized Abel transformation are of interest). In "curvilinear" domains and for irregular nets, it seems to be reasonable to include into consideration a metric tensor (its analog). This is also connected with the transition to equations (systems) with variable coefficients.

In the latter case, certain problems are encountered that refer to the possibility of simulating the degeneration (the degeneration of parts of the boundary inclusive) and a change of the type.

In Sec. 2, the isomorphism of d and δ homologies was considered in "good" domains, which are, for instance, "combinatorial manifolds" [30], but the conditions under which the isomorphism may be violated were not described.

In connection with the difficulties in the transition to "regular" boundary conditions for the "$d + \delta$"-system, encountered in Sec. 3, it would be interesting to study in detail the possibilities given by the use of the combinatorial analog of a double.

Our "nets" transferred to the surface of a cube or a torus will possess a special symmetry (admit of "groups of motions"). Symmetries of this kind can be used in the analysis of problems of a certain type (cf. [35]; Sec. 16 in [25]; Sec. 1, Ch. 5 in this book).

Various problems are connected with the hydrodynamic analogies of Sec. 4. There are two groups of problems of this kind. The first group refers to the study of the properties of the suggested equations in "large" domains of different structures and the second group is connected with the possibility of using the observed regularities for obtaining statements concerning continual objects (certain forms of a limiting process).

It would be interesting to find out whether our "combinatorial model" could be used for obtaining informative results that refer the analogs (approximations) of the Cauchy integral, analogs of the integrals of the Cauchy type, or with greater generality the analysis of certain classes of integral equations.

It is easy to notice that a considerable part of the problems indicated can be projected to continual objects of Ch. 2.

Chapter 4

Models in Quantum Physics

0. Introductory Remarks

The models discussed in this chapter differ from those studied in Ch. 3 not only formally but also in principle. We simulate here not so much as an analytic apparatus as methodological problems. A simplified scheme is used to discuss the idea of the mathematical formalism of physical theory.

This idea can roughly be understood as follows. The physical theory is associated with a certain mathematical **structure** which makes it possible to define **objects**, i.e., the subject of investigation, and some of their **characteristics**. The **task** of the theory is to find the characteristics of an object from its **input data**, that implicitly define it, on the basis of the laws following from the nature of the basic structure. It stands to reason that the task includes the finding of the indicated laws dictated by the structure, the elucidation of the general properties of the objects, the establishment of the relationship between the characteristics, etc.

The finiteness of the models used to illustrate the given scheme is brought about by the replacement of the space or the space-time continuum of "real" physics by a finite set.

The main scheme of Sec. 1 is the comparison, within the framework of the concepts being used, of the classical and the quantum description of the simplest object, namely, a "single mass point" or a "particle". We can show that the discrepancy between the indicated methods is less, as the more abstract and the less conventional initial "ordinary" concepts are formulated. In Sec. 2 we model the secondary quantification, i.e., the transition from a "particle" to a "pole".

It should be pointed out that in books on physics we can encounter the use of similar finite structures applied to the analysis of more special situations (say, [5, 67]).

The mathematical concepts used in this chapter differ essentially from those encountered in Chs. 2 and 3. The combinatorics that is akin to the homology theory is completely absent. Much space is given to the structure of a "homogeneous space", i.e., a set in which a group of mappings is defined.

In a continual case, the interaction of the two indicated structures (continuous and group-theoretic) considerably enriches the exposition, whereas their separation simplifies it.

It should be pointed out in conclusion that it would be interesting to compare the content of 1.2, 1.3 with the draft of quantum mechanics suggested in [27].

1. Models in Quantum Mechanics

1.1. Classical single mass point. Let us consider the process of formalization of the main concepts of classical mechanics by using the example of a single mass point of mass m moving in a straight line X under the action of the force $f(x)$, $x \in X$. This motion is described by the function $x(t)$ and is defined by the Newton law expressed, for a suitable choice of the units of measurement, by the differential equation

$$m\ddot{x} = f(x). \tag{1}$$

For the given initial conditions $x(t_0)$, $\dot{x}(t_0)$ the integration of the equation (1) makes it possible to find $x(t)$, and this allows us to solve every specific problem.

However, a different statement of the problem is also possible even for the elementary equation (1), namely, what are the general properties or laws inherent in **every** motion defined by Eq. (1)? It turns out that the transition to this aspect of the problem often serves as a stimulus for the introduction and analysis of concepts which find application in considerably more general and complicated situations. Let us use our example to follow the chains of definitions encountered in this way. We introduce the *potential*

$$U = -\int_0^x f(\xi)\, d\xi,$$

the *kinetic energy*

$$T = m\frac{\dot{x}^2}{2}$$

and the *total energy*

$$E = T + U.$$

PROPOSITION 1. *The conservation law*

$$\frac{dE}{dt} = 0$$

is valid for motion (1).

In order to prove this statement, it is sufficient to note that

$$\frac{dE}{dt} = \frac{\partial T}{\partial \dot{x}}\frac{d\dot{x}}{dt} + \frac{\partial U}{\partial x}\frac{dx}{dt} \quad \text{and} \quad \frac{\partial U}{\partial x} = -m\ddot{x}. \blacksquare$$

Alongside E, another important characteristic of motion (1) is the *Lagrange function*

$$L(x, \dot{x}) = T - U,$$

which makes it possible to introduce the *action*, namely, the functional

$$S = \int_{t_0}^{t_1} \mathrm{L}(x, \dot{x}), dt, \tag{2}$$

possessing the property of being valid.

PROPOSITION 2. *Eq.* (1) *is the Euler equation for functional* (2).

In order to prove this statement, it is sufficient to write out the relation

$$\frac{d}{dt}\Big(\frac{\partial L}{\partial \dot{x}}\Big) - \frac{\partial L}{\partial x} = 0. \blacksquare \tag{3}$$

It turns out that the notation of Eq. (1) in form (3) possesses a number of advantages, for example, the possibility of introducing "generalized coordinates" that include the possibility of considering, in the framework of the same scheme, nonholonomic systems that contain an explicit time dependence $L = L(x, \dot{x}, t)$, etc.

Moreover, the turning to functional (2) proves to be a prototype of the "Lagrangian formalism", i.e., of the starting point of various theories that cover the continuum and physical "fields" of different structures.

Another branch of mathematical constructions originating from Eq. (1) is connected with the notation of the total energy E in new variables:

$$E = \frac{p^2}{2m} + U(q) \equiv \mathbf{H}(p, q), \qquad q = x, \quad p = m\dot{x},$$

where $\mathbf{H}$ is the so-called *Hamilton's function* (*Hamiltonian*). Now we can write (1) as a system of equations

$$\frac{dq}{dt} = \frac{\partial \mathbf{H}}{\partial p}, \qquad \frac{\partial p}{\partial t} = -\frac{\partial \mathbf{H}}{\partial q} \tag{4}$$

and it turns out to be natural to consider the **phase** space $p, q \in M$ rather than the **configuration** space X. From the viewpoint of mathematics, we carry out, in this way, the transition from the **manifold** X to the **cotangent fibration** over X. In this event, Eqs. (4) define a vector field over M. (In the process of introduction of these definitions it is useful to "forget" that our X is nothing but $\mathbb{R}^1$.)

If now $\mathbf{F}(p, q)$ is a characteristic of motion, then

$$\frac{d\mathbf{F}}{dt} = \frac{\partial \mathbf{F}}{\partial q}\frac{dq}{dt} + \frac{\partial \mathbf{F}}{\partial p}\frac{dp}{dt} = \frac{\partial \mathbf{F}}{\partial q}\frac{\partial \mathbf{H}}{\partial p} - \frac{\partial \mathbf{F}}{\partial p}\frac{\partial \mathbf{H}}{\partial q} = [\mathbf{F}, \mathbf{H}], \tag{5}$$

where $[\mathbf{F}, \mathbf{H}]$ is the so-called Poisson bracket of the fields defined by the specification of $\mathbf{H}$ and $\mathbf{F}$.

REMARK. The introduction of the operation $[,]$ in the set of fields turns it into the **Lie algebra**.

The function $\mathbf{F}$ is known as the *motion integral* defined by $\mathbb{H}(p, q)$ if $\frac{d\mathbf{F}}{dt} = 0$. Thus relation (5) gives the following statement.

PROPOSITION 3. *The function* $\mathbf{F}$ *is the motion integral if and only if* $[\mathbf{F}, \mathbf{H}] = 0$. ∎

If we now write function (2) in the new variables, then

$$S = \int_{t_0}^{t} [-\mathbf{H} + p\dot{q}]\, dt \equiv \int_{t_0}^{t} \tilde{L}(q, p, \dot{q})\, dt,$$

and, again, Eqs. (4) turn out to be Euler equations for S. The transition from L to $\overline{L}$ is known as the *Legandre transformation.*

The various generalizations of the Hamilton formalism (which is based on the specification of a certain **Hamiltonian H**) competes successfully with the theories mentioned above which proceed from the definition of the **Lagrangian**.

REMARK. It should be added that a transition in the same line to the mechanics of the relativistic single mass point was considered in [69].

The following conclusion can be made from the given outline. The concepts used at different stages of describing a mechanical system (even the simplest one) can be highly abstract. In this case, the state of the system is regarded as a point of a certain set, namely, the space of states that possesses a certain additional structure; the motion of the system is a trajectory in the state space; the collection of all trajectories defines a one-parameter group of transformations of this space, etc.

This far-reaching inference must, in turn, serve as a motivation for the constructions that follow and that use a finite (consisting of a finite number of points) configuration space. As applied to the classical single mass point, the scheme that will be suggested does not advance beyond the "Newton law", but the "quantification" of this model described at the next stage is of a certain interest.

1.2. Model of a mass point. Suppose that Ω is a finite set of N elements to which is assigned the role of a physical space. The trajectories of the "mass point" lie in Ω, and in the sequel the "wave functions" will be functions over Ω. We define the principal object, a wandering point, as a function $q(t)$ of the discrete time $t \in \mathbb{Z}$:

$$q : \mathbb{Z} \to \Omega, \qquad t \mapsto q(t), \tag{6}$$

where $\mathbb{Z}$ is a set of integers. Let us suppose, furthermore, that we are given a transitive group $G : \ \Omega \to \Omega$ of the automorphisms of Ω. Thus, rejecting the structure of the differentiable manifold, we introduce a structure of a "homogeneous space". We suppose, in addition, that G is **simply transitive**, i.e., for any $x, y \in \Omega$ there exists a single element $g \in G$ such that $g : \ x \mapsto y$ (this corresponds to the consideration of the subgroup of translations in the group of all motions of $\mathbb{R}^3$).

Then object (6) is in a one-to-one correspondence, for any $t \in \mathbb{Z}$, with the element $p(t|q) \in G$ such that

$$p(t|q) : \ q(t) \mapsto q(t+1). \tag{7}$$

We can write relation (7) symbolically, as a relation

$$\Delta_t q(t) = p(t|q),$$

which is a definition of the map Δ_t which associates the trajectory $q(t)$ in Ω with the trajectory $p(t|q)$ in G.

Note that if we begin a similar construction with specifying a certain trajectory $p(t)$ in G, this will be equivalent to the specification of a **family** of trajectories in Ω, i.e., the individual object $q(t)$ will be defined only when the point $q(t_0) \in \Omega$ is additionally defined. At the next stage we have $q(t_0+1) = p(t_0)q(t_0)$, and so on (the reader is invited to compare the map p with classical momentum).

Let us now define $\Delta_t p$ setting

$$\Delta_t p(t) = p^{-1}(t)p(t+1) \in G.$$

The **principal postulate** ("Newton law of motion"). *The trajectory of the introduced object must be defined by the equation*

$$\Delta_t p(t|q) = \Phi[q(t)], \tag{8}$$

where $\Phi : \ \Omega \to G$ is a certain given map.

Thus we do not consider more general mappings $\Phi : \ \Omega \times \mathbb{Z} \to G$, which contain an explicit dependence on time, and the case of masses different from the unit mass. Just as in the classical case, problems connected with Eq. (8) may be different, ranging from the "integrating", i.e., defining $q(t)$ from the given Φ, $q(t_0)$, $p(t_0|q)$, to the establishment of certain general facts that refer to definite classes of motions. The simplest of these classes is a free motion corresponding to the choice of $\Phi : \ \Omega \to I$, where I is an identity of the group G. We shall consider an appropriate example in 1.4 in a somewhat more informative context.

1.3. Model of a quantified object. When passing to the "quantification", we shall restrict, as before, the constructions by the framework of the finite configuration space Ω introduced in 1.2. Now the definition of the analog of the classical elementary object considered in the preceding subsection requires additional constructions connected with the specification on Ω of a structure similar to that introduced in Sec. 2, Ch. 0.

We consider an *elementary object* to be the function

$$f : \Omega \to \mathbb{C}, \qquad x \mapsto f(x), \tag{9}$$

and introduce an inner product in the linear manifold of objects setting

$$(f, g) = N^{-1} \sum_{x \in \Omega} f(x)\overline{g}(x)$$

and turning it, in this way, into an N-dimensional Hilbert space H. We shall sometimes call the introduced object a "particle" emphasizing its kinship with a "mass point". We denote by $|f\rangle$ the element H normalized to unity and corresponding to the elementary object f:

$$|f\rangle = \frac{f}{|f, H|},$$

calling it the *state* of the elementary object (9). The object as such can be regarded in this case as the family $\{\varkappa f\}$, $\varkappa \in C$, of elementary objects (maps (9)) or as a ray of the Hilbert space H defined by the state $|f\rangle$.

We change the notation of the orthonormal basis $\{e_x\}$, introduced in Sec. 2, Ch. 0,

$$e_x(y) = \begin{cases} \sqrt{N}, & x = y, \\ 0, & x \neq y, \end{cases}$$

setting

$$e_x(y) \equiv |x(y)\rangle \equiv |x\rangle$$

and writing the corresponding decomposition as

$$f(x) = \sum_x f_x |x\rangle, \qquad f_x = \frac{f(x)}{\sqrt{N}}. \tag{10}$$

We introduce a special notation for the inner product of the states $|f\rangle$, $|g\rangle$, setting

$$\langle f|g\rangle = \frac{(f, g)}{|f, H|\,|g, H|}.$$

REMARK. We use the version of the so-called Dirac notations which are very expressive in a number of cases. We must bear in mind that in physical treaties that use these notations the bar (the sign of a complex conjugation) in an inner product is put over the **first** factor.

The given chain of definitions immediately yields the following statement.

PROPOSITION 4. *The sum* $\sum_1^n a_k|f_k\rangle$ *of the* n*-tuple of states* $|f_1\rangle, \dots, |f_n\rangle$ *with the complex coefficients* a_k*, normalized to unity, again defines a state.*

This statement expressed the so-called principle of superposition, one of the **postulates** of quantum mechanics.

Since we do not yet consider dynamics, we must refer the concept of a state that we have introduced to a certain fixed time momet t_0. From this point of view, object (6) can be considered to be a special case of a new object corresponding to the state $|f\rangle = |x\rangle$, $x = x(t_0)$. We call the transition from objects (6) to more general objects (9) *quantification.*

For the arbitrary state $|f\rangle$ the value $|f_x|^2$ in the representation

$$|f\rangle = \sum_k f_x|x\rangle$$

can be treated as the **probability** of discovering a particle at the point $x \in \Omega$ (at the given moment t_0).

From the formal point of view, the supposition of the transition from the consideration of a mass point to that of a "smeared" object, whose different parts fall in different places, is not worse. However, more customary is a probability interpretation which we shall adhere to.

We call the real functional, defined on a set of states, a *characteristic* of an object. Thus the functional

$$S_x|f\rangle = |\langle x|f\rangle|^2 = |f_x|^2$$

that we have considered, is a characteristic.

To carry out the requisite constructions, we must introduce some numerical coordinates in Ω, for instance, we can label the points $x, y, \dots$ by the numbers $0, \pm 1, \dots, \pm l$, $N = 2l + 1$ as we did in Sec. 2, Ch. 0. This will allow us to introduce such an important characteristic of a particle as the *mean value of the coordinate* in the state

$|f\rangle$:

$$\overline{X}|f\rangle = \sum_{-l}^{l} x|f_x|^2. \tag{11}$$

It is now useful to note that this functional belongs to one of the most important standard classes of real functionals over a Hilbert space, namely, to the class of functional generated by a certain Hermite operator $A: H \to H$, according to the formula

$$\overline{A}|f\rangle = \langle f|A|f\rangle \equiv (f, Af). \tag{12}$$

The standard way of notation, used in the middle term, emphasizes the self-conjugacy (Hermitian property) of the operator A, and the parentheses in the last term mean that the element Af must not necessarily be normalized to unity. In this context the operator A itself is called an *observable.*

For (11) the part of A is played by the operator X of "multiplication by the coordinate", defined by the relation

$$X|f\rangle = \sum_x x f_x |x\rangle. \tag{13}$$

Correspondingly, relation (12) gives the mean value of the arbitrary observable A.

It follows from (13) that the basis $\{|x\rangle\}$ is the basis of the **eigenvectors** of the operator X and (10) is the notation of the decomposition of the state $|f\rangle$ into components with respect to these eigenvectors. The resolution of $|f\rangle$ into components with respect to the eigenvectors of the vector A plays a similar part when we calculate the mean values of the operator A defined by (12); or, in a more general case, when we calculate the probability of discovering, in the state $|f\rangle$, the values of the observable which fall in the given interval.

Before considering this fact in greater detail, let us take into account the following important remark. Just as in Sec. 2, Ch. 0, the existence of a basis of eigenvectors of the operator X is a specific "convenience" of a finite model, namely, when we pass to a continual case, the element $|x\rangle$ becomes a generalized function that **does not belong** to the space H.

Thus, suppose that A is an observable, $\{|u_k\rangle\}$ is a complete orthonormal system of eigenvectors of the operator A, $|f\rangle$ is a certain state, and

$$A|u_k\rangle = \lambda_k|u_k\rangle, \qquad |f\rangle = \sum f_k|u_k\rangle.$$

Then the probability of finding in this state the values of the observable A lying in the interval $\alpha \in [a, b]$ is given by the relation

$$P_\alpha = \sum_{k \in \omega(\alpha)} |f_k|^2, \qquad k \in \omega(\alpha) \quad \text{if} \quad \lambda_k \in \alpha. \tag{14}$$

Understanding the integral as the corresponding sum, we can give (14) the form of the classical formula

$$P_\alpha = \Big\langle f \Big| \int_{\lambda \in \alpha} f \, dE_\lambda(A) \Big\rangle.$$

Returning to the mean values, let us define the most significant characteristic of the "observable–state" pair, the so-called *dispersion*: i.e., the mean square deviation of the values of the observable A in the state $|f\rangle$ from the mean $\overline{A}|f\rangle$. Introducing the notation $A_\delta = A - I\overline{A}$, where I is an identity operator, and noting that the operator A_δ is again self-conjugate, we can write

$$\delta^2(A, f) = \langle f | A_\delta^2 | f \rangle = |A_\delta f, H|^2.$$

In particular, taking as A the "operator of the coordinate" X considered above, we can naturally consider the quantity $\delta^2(X, f)$ to be the "measure of nonclassicality" of the object. It is easy to calculate that $\delta^2(X, f)$ attains its minimum on the "classical states" $|f\rangle = |x\rangle$ ($\delta^2(X, x) = 0$) and its maximum on the states that do not contain any information concerning the position of the particle, i.e., $|f\rangle = N^{-1/2}$ for any $x \in \Omega$.

Dispersion also enters into the mathematical formulation of the Heisenberg **"principle of indeterminacy"**.

PROPOSITION 5. *For any pair of self-conjugate operators A, B : $H \to H$ and the arbitrary state $|f\rangle$, we have*

$$\delta(A, f)\delta(B, f) \geq \frac{1}{2} |\langle f | [A, B] | f \rangle|, \tag{15}$$

where $[A, B]$ is the commutator of the operators A, B (which is a self-conjugate operator).

PROOF. Noting that $[A - \alpha, B - \beta] = [A, B]$ for any constants α, β, we can write

$$|\langle f | [A, B] | f \rangle| = |\langle f | [A_\delta, B_\delta] | f \rangle| = |\langle B_\delta f | A_\delta f \rangle - \langle A_\delta f | B_\delta f \rangle| =$$

$$2|\mathrm{Im}\,\langle A_\delta f|B_\delta f\rangle| \le 2|\langle A_\delta f|B_\delta f\rangle| \le 2|A_\delta f, H|\,|B_\delta f, H|. \blacksquare$$

Inequality (15) is usually connected with the remark to the point that the simultaneous minimization of $\delta(A, f)$, $\delta(B, f)$ is impossible for a nonclassical object for noncommuting A, B (hence the "indeterminacy principle").

It should also be pointed out that the quantity $|\langle f|g\rangle|$ is often regarded as the **measure of proximity** of the states $|f\rangle$, $|g\rangle$ connected with the probability of a change from one state to the other.

At the next stage we turn to the **dynamics**, i.e., to the description of the change of the state $\langle f|$ with time. As before, we assume it to be discrete. Now the analog of (6) is the mapping

$$F : \mathbb{Z} \to H, \qquad t \mapsto |f(t)\rangle$$

of the set of integers into a unit sphere of the space H. Instead of (7) we must now have

$$U(t|f) : \ |f(t)\rangle \mapsto |f(t+1)\rangle, \tag{16}$$

where $U : \ H \to H$ is a certain unitary operator and $\Delta_t f(t) = U(t|f)$. Then

$$\Delta_t U(t) = U^{-1}(t)U(t+1) \in \mathfrak{G},$$

where $\mathfrak{G}$ is a group of unitary transformations of H, and a natural analog of the dynamic postulate (8) should be a law of the form

$$\Delta_t U(t|f) = \Phi(|f(t)\rangle),$$

where $\Phi : \ H \to \mathfrak{G}$ is a given map. However, in actual fact, a different relation is chosen as the basic dynamic postulate of quantum mechanics. We shall construct its analog.

Since H is a complex linear space, the difference

$$\partial_t|f(t)\rangle = |f(t+1)\rangle - |f(t)\rangle = [U(t|f) - I]|f(t)\rangle, \tag{17}$$

is defined which had no sense in the classical variant of our model.

If we now note that for any unitary operator U we have the representation $U = e^{-i\mathbf{H}}$, where $\mathbf{H}$ is a Hermitian (we have chosen the sign before $i\mathbf{H}$ by convention), and make the following additional assumptions:

1. In (17) we can replace the difference $U - I = e^{i\mathbf{H}} - I$ by the principal term, namely, $i\mathbf{H}$ (the assumption of the "smallness' of the steps with respect to t),

2. The operator $\mathbf{H}$ does not depend explicitly on time and on the element $|f\rangle$,

then (17) will turn into the "Schrödinger equation":

$$\partial_t |f(t)\rangle = -i\mathbf{H}|f(t)\rangle, \tag{18}$$

which makes it possible to formulate the following postulate.

THE FUNDAMENTAL POSTULATE. *The "trajectory" of the quantified object is defined by Eq.* (18), *where* $\mathbf{H}: H \to H$ *is a certain given Hermitian operator.*

It is an act of mastership to choose the operator $\mathbf{H}$ in the process of analyzing a physical situation (just as the choice of dynamic equations in nontrivial problems of classical mechanics). Here a source of inspiration is the so-called principle of correspondence, namely, a recipe which allows us to use the Hamiltonian of a classical system in order to write the operator $\mathbf{H}$ of the "corresponding" quantum system. The mathematical justification of the principle of correspondence is the possibility of obtaining, in the limit, the classical mechanics from the quantum mechanics when a certain parameter (not participating in our constructions), that characterizes the measurement scales, tends to zero. However, a thorough analysis shows that the corresponding limiting process is, in general, not unambiguous [32] and must be carried out with the use of a special technique.

In the next subsection we shall be able (by turning Ω into a "homogeneous space") to get a reasonable answer to the following question: "What must the operator $\mathbf{H}$ be in (18) if this equation defines the trajectory of a classical free mass point?" For the time being, we shall consider a transition to a different method of describing a change in time of a quantified system which uses the so-called Heisenberg approach.

For our state $|f(t)\rangle$ varying with time we shall write the mean value of the observable A at the moment $t+1$:

$$\overline{A}|f(t+1)\rangle = \langle f(t+1)|A|f(t+1)\rangle = \langle Uf(t)|A|Uf(t)\rangle =$$

$$\langle f(t)|U^{-1}AU|f(t)\rangle.$$

We can now introduce the dependence of the **observable** A on time by setting $A(t+1) = U^{-1}AU$, and calculate $\partial_t A = A(t+1) - A(t)$. Then, restricting our consideration to the linear terms of the corresponding decompositions, we obtain

$$\partial_t A = U^{-1}AU - A \approx (I + i\mathbf{H})A(I - i\mathbf{H}) - A \approx i(\mathbf{H}A - A\mathbf{H})$$

or

$$\partial_t A = i[\mathbf{H}, A],$$

and this gives the description of the dynamics of the system defined by the operator $\mathbf{H}$ via the variation of the observable with time.

1.4. The correspondence principle. Let us return to the model of a classical mass point from 1.2 specifying the choice of G and, at the same time, introducing numerical coordinates in the "homogeneous space" Ω as we did in 1.3. In order to avoid numerous references to the corresponding places in the text, we shall repeat below a considerable part from Sec. 2, Ch. 0.

Thus, let us suppose that the group G is a finite cyclic group with a generator τ which acts in the set Ω whose points are labeled by the numbers $0, \pm 1, \dots, \pm l$, according to the rule

$$\tau : x \mapsto x + 1, \quad x \neq l, \qquad \tau : l \to -l, \quad 2l + 1 = N.$$

If we now set $\Phi = 0$ in (8) (using the customary notation for the unit element of the Abelian group), then we get

$$p(t) = \text{const} = \tau^m,$$

with a certain fixed value of m defined by the "initial value" $p(t_0)$. Then $q(t+1) = \tau^m q(t)$, i.e., the position of the point at the moment $t+1$ results from the "shift by m units" of its position at the moment t.

In terms of 1.3, in accordance with remarks made there on the state corresponding to a classical particle, the situation we have described can be expressed by the relation

$$|x(t + 1)\rangle = \hat{\tau}^m |x(t)\rangle, \tag{19}$$

where the unitary operator $\hat{\tau} : H \to H$ is a standard representation of the shift $\tau : \Omega \to \Omega$, defined on the elements $|f\rangle$ by the law $\hat{\tau} f(x) = f(\tau^{-1}x)$, where we used decomposition (10) for $|f\rangle$. The motion law (19) can obviously be generalized to the arbitrary state $|f(t)\rangle \in H$:

$$|f(t + 1)\rangle = \hat{\tau}^m |f(t)\rangle. \tag{20}$$

We can now formulate the question concerning the correspondence of the classical and the quantum model constructed in this way: "What must the operator $\mathbf{H}$ be in the representation $U = e^{-i\mathbf{H}}$ if the motion law written in the form (16) has the form (20)?"

However, before answering the question stated in this form, let us consider its other aspects, returning to relation (13). It shows that we can use, in particular, the operator X, or even the operator of multiplication by the arbitrary real function $\varphi(x)$, as the operator H. Then (turning to a continual variant) we get

$$Hf = \varphi(x)f, \qquad \frac{\partial f}{\partial t} = i\varphi(x)f,$$

or $f = c(x)e^{it\varphi(x)}$. If we consider $c(x)$ to be a smooth finite function, then this amplitude f describes the simplest oscillation process.

It is natural to consider an immediate generalization of the result obtained setting $f = e^{-iS(x,t)}$ (the "Feynman amplitude"; the minus in the exponent is conventional). Then the equation assumes the form $\frac{\partial f}{\partial t} = -i\frac{\partial S}{\partial t}f$ and, respectively, $Hf = -\frac{\partial S}{\partial t}f$. The resulting symbolic equality $\frac{\partial S}{\partial t} = -H$ can be regarded as an analog of the classical Hamilton–Jacobi equation in which S is an action and H is a Hamiltonian. Correspondences of this kind are often discussed in courses of quantum mechanics.

Let us return to relation (20). In order to get the answer we are interested in, we must now use the decomposition of $|f\rangle$ into components with respect to the eigenvectors of the shift operator $\hat{\tau}$. The corresponding basis H is defined by the elementary "periodic functions", i.e., by a collection of roots of unity:

$$\eta_k|x = e^{\nu kx}, \qquad \nu = N^{-1}2\pi i, \quad k = 0, \pm 1, \dots, \pm l. \tag{21}$$

Then

$$\hat{\tau}|\eta_k\rangle = \alpha_k|\eta_k\rangle, \qquad \alpha_k = e^{-\nu k},$$

and, simultaneously,

$$\langle\eta_k|\eta_j\rangle \doteq \delta_{kj},$$

i.e., the chosen basis is orthonormal. Correspondingly, we have now two representations,

$$|f\rangle = \sum_x f_x|x\rangle = \sum_k \tilde{f}_k|\eta_k\rangle,$$

for the element $|f\rangle \in H$, where the relationship between the coefficients is established by the relation

$$\tilde{f}_k = \langle \eta_k | f \rangle = \sum_x f_x \langle \eta_k | x \rangle = N^{-1/2} \sum_x f_x e^{\nu k x},$$

since

$$\langle \eta_k | x \rangle = N^{-1} \sum_y \eta_k(y) e_x(y) = N^{-1/2} e^{\nu k x}.$$

The reverse transition has the form

$$f_x = N^{-1/2} \sum_k \tilde{f}_k e^{-\nu k x}. \tag{22}$$

If we use the symbolics employed in the Fourier transformation theory, then the relationship between f and $\overline{f}$ can be expressed by the relations

$$\tilde{f}_k = F f_x, \qquad f_x = F^{-1} \tilde{f}_k.$$

REMARK. It should be noted that this time, as distinct from Ch. 0, the part of the values $f(x)$ of function (1), Sec. 2, Ch. 0, is now played by the coefficients f_x of the decomposition of $|f\rangle$ into components with respect to the natural base vectors.

When using basis (21) for describing quantified objects, it is customary to regard the values of the parameter $k = 0, \pm 1, \dots, \pm l$ as numerical coordinates on a certain set $\tilde{\Omega}$ and consider over $\tilde{\Omega}$ the complex Hilbert space $\tilde{H}$ in which the vectors $|k\rangle$ play the same part as $|x\rangle$ plays in H. Then we say that the state $|\tilde{f}\rangle \in \tilde{H}$ is given in a "momentum representation".

Let us now find out the form of the operator $\mathbf{H}$ for motion (20). If

$$|f(t)\rangle = \sum_k \tilde{f}_k(t) |\eta_k\rangle,$$

then

$$U|f(t)\rangle = \hat{\tau}^m |f(t)\rangle = \sum_k \tilde{f}_k(t) \alpha_k^m |\eta_k\rangle = \sum_k \tilde{f}_k(t) e^{-\nu m k} |\eta_k\rangle. \tag{23}$$

However, if $\{\varphi_k\}$ is the basis of the eigenvectors of a certain operator A and the action of A can be represented as $A\varphi_k = e^{\lambda_k}\varphi_k$ (the eigenvalues of A are written in the form of exponents), then, in accordance with the rules of definition of functions of operators, we can set $A = e^B$, where the action of B is defined by the relation

$B\varphi_k = \lambda_k \varphi_k$. Applying these arguments to (23) and considering the representation $U = e^{-i\mathbf{H}}$, we obtain $\mathbf{H}|\eta_k\rangle = \alpha_k|\eta_k\rangle$, $\alpha = N^{-1}2\pi m$.

PROPOSITION 6. *In the momentum representation the operator* $\mathbf{H}$ *has the form* αK, *where* $K : \tilde{H} \to \tilde{H}$ *is an operator of multiplication by the variable* k.

Indeed, if $|\tilde{f}\rangle = \sum_k \tilde{f}_k|k\rangle$, $K|\tilde{f}\rangle = \sum_k k\tilde{f}_k|k\rangle$, then $H|\tilde{f}\rangle = \alpha \sum_k k\tilde{f}_k|k\rangle$. ■

We can now find out to what the operator $\mathbf{H}$ corresponds in the "coordinate" representation.

PROPOSITION 7. *The operator* $D : H \to H$, *defined by the relation* $Df = F^{-1}K\tilde{f}$, *has the form*

$$(Df)_x = \nu^{-1}[f_{x-1} - f_x] + \varepsilon_\nu(f), \tag{24}$$

where the remainder ε_ν *is defined below.*

PROOF. According to (22),

$$(Df)_x = N^{-1/2} \sum_k k\tilde{f}_x e^{-\nu kx}.$$

Using the representation $k = \nu^{-1}(e^{\nu k} - 1) + \tilde{\varepsilon}_\nu(k)$, we can write

$$(Df)_x = \nu^{-1}N^{-1/2} \sum_k (e^{-\nu k(x-1)} - e^{-\nu kx})\tilde{f}_k + \varepsilon_\nu(f),$$

where $\varepsilon_\nu(f) = F^{-1}(\tilde{f}\tilde{\varepsilon}_\nu)$, and this gives (24). ■

Thus, "in the limit", the operator D would correspond to the differentiation operator (multiplied by i). The constant α, that appears in the definition of $\mathbf{H}$, is a characteristic of the system which defines the relationship between the "momentum" and the "velocity".

Considering Proposition 5, we can note that $[D, X] \neq 0$, hence the impossibility of a simultaneous exact definition of the "coordinate" and the "momentum" of the quantified system.

Let us now consider some discrepancies between our model and the standard formalism of quantum mechanics. The result formulated in Proposition 6 differs from the prescription of the classical correspondence principle which states that the operator H for free motion must have the form $\frac{1}{2}K^2$ (for $\alpha = 1$), and this corresponds to the form of the Hamiltonian (equal to $\frac{1}{2}p^2$) for the free motion of a mass point (see 1.1).

In our approach, the continual analog of the Schrödinger equation would be the equation (for the system under consideration)

$$iD_t f = (viD_x)f, \tag{25}$$

where v is a real constant and iD_x is a self-conjugate operator generated by the differentiation operation. Writing (25) simply in the form

$$\frac{\partial f}{\partial t} - v\frac{\partial f}{\partial x} = 0,$$

we get, as the general solution, the wave $f(vt + x)$ that spreads without any distortion of the form.

Wave functions of this type are widely used as the zero approximation, say, in dissipation problems, but additional efforts are sometimes required to obtain them from the above-mentioned standard Hamiltonian (cf. [9, p. 62]).

It is natural to add that when we have a multiplication operator and the operator $i\frac{\partial}{\partial x}$ in our arsenal, the choice of arguments that would lead to more complicated "Hamiltonians" (to the classical $Hu = \frac{\partial^2 u}{\partial x^2} + V(x)u$, for instance) can be left to the reader.

In accordance with what we have said, let us consider one more circumstance. In our considerations we completely ignored the requirement (the postulate) of positiveness of the Hermitian operator $\mathbf{H}$ ("positiveness of energy"). This requirement was implied by the choice of the minus sign in $\exp(-i\mathbf{H})$. One of the simplest techniques of transition from the operator $\mathbf{H}$, generated by K, to a positive operator is, from a purely theoretic point of view, the replacement of K by $(K^2+\mu^2)^{1/2}$ (or simply by $K^2+\mu^2$), where the constant $\mu^2 > 0$ will define the lower bound of the spectrum. It turns out that the corresponding Hamiltonian $\mathbf{H}_\mu : H \to H$ is actually used as one of the standard "free Hamiltonians" in models of the relativistic field theory with one space variable.

2. Models in Quantum Field Theory

2.0. Preliminary remarks. Bearing in mind that we use, as before, a perfectly simple model of a physical space, we shall try to analyze a number of constructions which are widely used in theoretic physics but which hardly yield to formal mathematical interpretation. Constructions of the type that we suggest can possibly contribute, to a certain degree, to filling in the gap between the works which are mathematically perfect but poorly suitable for the elucidation of the general situation and encyclopedic guides ([57], for example) since they require the skill of operating with objects which do not have exact definitions.

The initial setups and a developed plan of the section are given in 2.1.

2.1. The main features of transition to the field theory. The following is a naive definition of the field which is prompted by our intuition. The *field* Φ (over the set M) is an object for which, for an arbitrary choice of $x \in M$, the following question is considered to be meaningful: "What is the characteristic of Φ at the point x?"

From this point of view, the states $|f\rangle$ from Sec. 1 could be regarded as an example of a field ("the amplitude of a scalar field") provided that, in accordance with other postulates, we would have at our disposal the operator $\Phi(x)$ (dependent on x as on a parameter) such that the number $\langle f|\Phi(x)|f\rangle$ would give the required characteristic at the point x. In a finite-dimensional model it is difficult to keep from suggesting the consideration, first of all, of the operator "of the density of a particle at the point x", for which

$$\langle f|\Phi(x)|f\rangle = |f(x)|^2.$$

Nevertheless, as we shall see later, the construction of a theory rich in content requires the use of considerably more complicated constructions. At the same time, the idea of using the element $|f\rangle$ as the amplitude of the simplest field (scalar, one-particle) proves to be a lucky one.

Summing up what was said, we agree to regard the *introduction into consideration of a special class of observables–operators, dependent on the coordinates of the points of the space-time as one parameter*, as the **first** significant feature of the quantum field theory (QFT).

Note that we understand a field in QFT to be some collection $\{\Phi(x,t)\}$ of field operators and not the amplitude of $|f\rangle$. This is conventional and is also connected with a primary use of the Heisenberg representation in dynamics (see the end of 1.3).

It is interesting to point out that the first indicated feature is associated with the interpretation of QFT as the next step in the development of the ideas concerning the concept of a state, namely, a state-number (classical mechanics), a state-function (quantum mechanics), and a state-operator (QFT). The first step is known as **quantification**, the second as the **second quantification**. These steps can also be interpreted at the next "story" corresponding to the "building of the observables": a function $\to$ an operator $\to$ an operator function.

The **second** main feature of QFT can be defined as the *requirement of the creation of a mathematical apparatus that would make it possible to describe the processes of birth, annihilation, and mutual transformation of "elementary particles"*. We shall consider the specific structures that make it possible to meet this requirement in 2.2.

Finally, the **third** main feature is the *necessity of making the processes, given by the apparatus of QFT, consistent with the postulates of the special theory of relativity* which contain the relationship between the mass and the energy and the requirement of "relativistic covariance", i.e., the compatibility of the laws of transformation of the quantities being considered with the idea of the equivalence of the coordinate systems connected by the Lorentz transformation.

The consideration of this feature on a finite model leads to the introduction, on M, of the structure of a vector space over a **finite field** $\mathcal{F}$ and to the study of the model of the Poincaré group. Subsection 2.3 is devoted to the corresponding constructions.

Subsection 2.4 is a certain "synthesis" of the two preceding subsections. Subsection 2.5 contains a scheme of the construction of dynamics on the basis of perturbation theory.

2.2. The Fock space. In QFT the amplitudes of a field are defined, as a rule, in the momentum representation. We shall also carry out the constructions that we shall need in this representation. It is convenient to change somewhat the notations used in Sec. 1.

Suppose that $\tilde{M}$ is a momentum space, i.e., a finite set, whose N points have numerical coordinates $k = 0, \pm 1, \ldots, \pm l$, $N = 2l+1$, and $|f\rangle$ is a one-particle amplitude which can be regarded as a complex function over $\tilde{M}$ or as an element $|f\rangle = \sum f(k)|k\rangle$, $\| \, |f\rangle \| = 1$, of an N-dimensional vector space with the orthonormal basis $\{|k\rangle\}$ of δ-functions (as distinct from Sec. 1, we shall not put a tilde over f). Then the n-particle state must be described either by the collection $\{|f_i\rangle\}$ of amplitudes, $i = 1, \ldots, n$, or by the function over $\tilde{M} \times \ldots \times \tilde{M}$ (n factors). We choose the latter. The corresponding amplitude will be an element of the tensor product $\tilde{H}^n = \tilde{H} \otimes \ldots \otimes \tilde{H}$ (n factors), i.e., a vector of the space of dimension N^n with an orthonormal basis, consisting of the vectors

$$|k_1\rangle \otimes \ldots \otimes |k_n\rangle, \qquad k_s = 0, \pm 1, \ldots, \pm l. \tag{1}$$

This construction must be completed by the following postulate.

POSTULATE 1. *The amplitudes of real physical n-particle states belong either to a symmetric* (S^n) *or to a skew-symmetric* (Λ^n) *tensor algebra over* $\tilde{H}^n$ (cf. 3.6, Ch. 1).

In what follows we shall restrict the discussion to a **symmetric** case of the so-called scalar bosons and the corresponding fields. Then the basis in S^n must consist of the sum of vectors of the form (1) taken over various permutations of the n-tuple $(k_1, \ldots, k_n)$. In accordance with our conventions, I shall give an explicit description of these bases in the case of $N = 3$. (The combinatorics used in the general case can be found in [57].)

We denote the vectors $|-1\rangle$, $|0\rangle$, $|1\rangle$ as e_1, e_2, e_3, respectively. Then, for $n = 1, 2, 3$ we have

$$S^1 = \{e_i\},$$

$$S^2 = \{e_i \otimes e_i;\ 2^{-1/2}(e_i \otimes e_j + e_j \otimes e_i)\},$$

$$S^3 = \{e_i \otimes e_i \otimes e_i;\ 3^{-1/2}(e_i \otimes e_i \otimes e_j + e_i \otimes e_j \otimes e_i + e_j \otimes e_i \otimes e_i);$$

$$6^{-1/2}(e_i \otimes e_j \otimes e_l + (5 \text{ terms given by different permutations}))\}.$$

The subscripts i, j, l take on the values 1, 2, 3 and are assumed to be pairwise distinct. The presence of numerical factors is justified by the requirement of the orthonormality.

If we accept conventional polynomial notations, then the base elements written out can be represented as

$$\{\xi_i\}, \qquad \{\xi_i^2, \xi_i\xi_j\}, \qquad \{\xi_i^3, \xi_i^2\xi_j, \xi_i\xi_j\xi_l\}.$$

Then, for an arbitrary n $(N = 3)$, the element S^n admits of a representation of the form

$$|f_{(n)}\rangle = \sum_{\alpha_1+\alpha_2+\alpha_3=n} f_{\alpha_1\alpha_2\alpha_3}\xi_1^{\alpha_1}\xi_2^{\alpha_2}\xi_3^{\alpha_3} \equiv \sum_{|\alpha|=n} f_\alpha \xi^\alpha,$$

where $\sum_\alpha |f_\alpha|^2 = 1$ and $|f_\alpha|^2$ for $\alpha = (\alpha_1, \alpha_2, \alpha_3)$ can be regarded as the probability of discovering, in the given state $|f\rangle$ (multiparticle in this case), α_1 particles with momentum -1, α_2 particles with momentum 0, and α_3 particles with momentum 1.

Now, in accordance with the second feature of QFT, in order to be able to consider processes with varying number of particles, we must introduce a space S which is the direct sum of all S^n. In this event, as will be clear later, it is natural to include into this sum a term S^0, which is a one-dimensional space with a base element ξ_0 corresponding to the "particle-free state", i.e., *vacuum.* Thus we have

$$S = \overset{\infty}{\underset{0}{\oplus}} S^n, \qquad \|f\| = \sum_0^\infty \|f_{(n)}\|, \tag{2}$$

where we have omitted the vector symbol $|\ldots\rangle$ in f. Thus the elements of S are finite or infinite sums normalized to unity. The space S that we have introduced is known as the *Fock nonrelativistic space.*

It is known how tedious it is to consider spaces of type (2), and for infinite-dimensional S^n at that, as we have in a real situation. However, the natural attempt to restrict the consideration to a certain number $n \le n_0$ in processes that admit of a variation of the number of particles usually leads to the violation of the internal harmony of constructions. A reasonable "cutting with respect to the number of particles" can be performed only in the framework of perturbation theory since it leads to the consideration of the Feynman amplitudes.

Let us consider **operators** in the Fock spaces. We begin with a remark that in the general theory of Hilbert spaces, the most important kinds of operators are the so-called *shift* operators: $\tau: \; e_j \to e_{j+1}$, and the *generalized shift* operators $a\tau$, where a is a complex number ($\{e_j\}$ are base elements). The related operators, the so-called *birth* and *annihilation* operators, serve as the basis for the important formalism of QFT.

The operator A_j of *annihilation* of one particle with the value of the momentum corresponding to the base element ξ_j is defined (in the framework of our model, i.e., $N = 3$) for $|\alpha| \geq 0$ by the relation

$$A_j \xi_i^{\alpha_i} \xi_j^{\alpha_j} \xi_l^{\alpha_l} = \sqrt{\alpha_j} \xi_i^{\alpha_i} \xi_j^{\alpha_j - 1} \xi_l^{\alpha_l}, \tag{3}$$

where the right-hand side is set equal to zero for $\alpha_j = 0$ and, at the same time, $A_j \xi_0 = 0$ for any j. The numerical factor $\sqrt{\alpha_j}$ entering into (3) leads to the **unboundedness** of the operator $A_j : S \to S$ even in the framework of our model. The introduction of this factor, which causes many difficulties, is necessary, as we shall see later, for obtaining the operator of the **number of particles**; this, in turn, is a constituent part of the operator of **energy**, namely, a "free Hamiltonian", which is a starting point for constructing dynamics.

It is easy to verify that

$$A_j^* \xi_i^{\alpha_i} \xi_j^{\alpha_j} \xi_l^{\alpha_l} = \sqrt{\alpha_j + 1} \xi_i^{\alpha_i} \xi_j^{\alpha_j + 1} \xi_l^{\alpha_l}$$

is a conjugate of A_j in S. The corresponding definitions for an arbitrary N are obvious, but the notation becomes cumbersome. The self-conjugate operator

$$A_j^* A_j = \alpha_j \tag{4}$$

is an operator of the multiplication of an element of the basis by the number α_j and is called the operator of the *number of particles* (in a state with the value $k = k(j)$ of the momentum). Since, at the same time, $A_j A_j^* = \alpha_j + 1$ (in the notation similar to (4)), the commutator $[A_j, A_j^*]$ gives an identity operator. For $i \neq j$ all operators A_i, A_j commute. The totality of the remarks that we have made is usually written by means of the conventional formula

$$[A_i, A_j^*] = \delta(i - j).$$

The operators

$$\sum_j A_j^* A_j, \qquad \sum_j |k(j)| A_j^* A_j,$$

are known as the operators of the *complete number of particles* and the *energy* operator. In this case, the energy is usually considered to be "relativistic", i.e., instead of $|k|$ we have the factor $(k^2 + m^2)^{1/2}$, but we shall be able to give it some sense only in 2.4.

As a result, we have examples of families of operators dependent on a parameter, namely, on a coordinate (a momentum coordinate in this case), i.e., the first feature of QFT was reflected in the arguments that we have given (when we constructed the Fock space, we began with the second feature). Before passing to the third feature, let us consider the simple dynamic example connected with the redistribution of the number of particles.

We shall introduce the time dependence in the approach that we use simply as the dependence of the amplitudes, i.e., the elements of S, on the additional parameter t. A primitivism of this kind is seldom used since the requirement of the Lorentz covariance is violated in it.

Alongside (4), one of the simplest self-conjugate operators, defined in terms of A_j, is the operator

$$A_j + A_j^*. \tag{5}$$

Operators of this kind are widely used in QFT. We shall use (5) to construct a model of the Hamiltonian. We set

$$F = \sum_j (A_j + A_j^*), \qquad \mathbf{H} = \pi_0 F + F\pi_0,$$

where $\pi_0 : S \to S_0$ is a projection operator. Next we introduce a further simplification setting $N = 2$, i.e., assuming that the basis in S^1 is simply defined by the pair (ξ_1, ξ_2). Then

$$\mathbf{H}\xi_0 = \xi_1 + \xi_2, \qquad \mathbf{H}\xi_2 = \mathbf{H}\xi_2 = \xi_0,$$

and the other base vectors in S are annihilated by the operator $\mathbf{H}$. The vectors

$$h_0 = (0, 2^{-1/2}, -2^{-1/2}), \quad h_1 = (-2^{-1/2}, 1/2, 1/2),$$
$$h_3 = (2^{-1/2}, 1/2, 1/2)$$

are orthonormal eigenvectors for $\mathbf{H}$ that belong to the eigenvalues 0, ± 2 and that give the basis $S^0 \oplus S^1$. If we suppose that $\mathbf{H}$ is the Hamiltonian that defines the Schrödinger dynamics, then

$$f(t) = \exp(-it\mathbf{H})f(0),$$

where the "time" t is a certain parameter (say, an integer parameter). For the initial states $f(0) = h_s$, where h_s is one of the base vectors

$(s = 0, 1, 2)$, we have $f(t) = \exp(-it\lambda_s)h_s$, where λ_s is real, i.e., the corresponding states are stationary. Now if, for instance, $f(0) = \xi_1 = \frac{1}{2}(2^{1/2}h_0 + h_1 + h_2)$, then

$$f(t) = \exp(-it\mathbf{H})xi_1 = \frac{1}{2}\{2^{1/2}h_0 + \exp(i2^{1/2}t)h_1 + \exp(-i2^{1/2}t)h_2\} =$$

$$\Big(-\frac{1}{2}\sin(2^{1/2}t)\Big)\xi_0 + f_1(t)\xi_1 + f_2(t)\xi_2,$$

i.e., the "annihilation of a particle" can occur with a nonzero probability.

The use of simple Hamiltonians defined directly in terms of A_j, A_j^* is the main idea of the so-called **Lie model** [57], which makes it possible to analyze a number of qualitative aspects of the processes of transformation of particles, but the complete absence of space representations in it makes it "poor" and nonrealistic. We shall consider more realistic field operators in 2.4.

2.3. The Minkowski space and axiomatics. The difficulties connected with carrying out the constructions, similar to those given in 2.2 and considered in 2.4 and 2.5, in a continual case led to the appearance in QFT of a line of investigation known as the axiomatic approach. From the point of view of the axiomatic approach, the *field* $\Phi(\mathbf{x})$ (scalar) is the map of the space-time (the Minkowski space) M

$$\Phi : M \to \mathfrak{U}(H), \qquad \mathbf{x} \mapsto \Phi(x)$$

in the algebra of the Hermitian operators $\mathfrak{U}$ over a certain complex Hilbert space H which plays the part of the space of amplitudes (the part of our space S). However, the nature of H is not defined concretely.

In order to be able to reflect the aims of this line of investigation in our models, we must construct a finite analog of the relativistic space-time (two-dimensional at least). This analog M^2 must be a two-dimensional vector space with coordinates (t, x) and a pairing (indefinite inner product) that defines the Lorentz metric. For M^2 to be a finite set, the field of scalars must be **finite** in contrast to the classical case described in Sec. 3, Ch. 1. (The construction will give the model of the third feature of QFT.)

Although the concepts that we use here are elementary, we have to go a little beyond the framework of the "basic structures" of Ch. 1

and refer the reader to [26, 29]. It should be pointed out that the constructions we give are based on the article [67].

Thus we introduce a finite field $\mathcal{F}_p = \mathbb{Z}/p\mathbb{Z}$ setting $p = 7$. It will be clear in 2.4 that it exactly corresponds to the choice of $N = 3$ for the described S. The elements of $\mathcal{F}$ (we omit the index 7) can be labeled by the numbers

$$0, \pm 1, \pm 2, \pm 4. \tag{6}$$

The operations are defined in an ordinary way, i.e., as the addition and multiplication mod 7. The nonzero elements of $\mathcal{F}$ are divided into two groups, namely, a group of squares $\mathcal{F}_+ = \{1, 2, 4\}$ and a group of nonsquares $\mathcal{F}_- = \{-1, -2, -4\}$ (i.e., elements for which the Legendre symbol is equal to $+1$ or -1, [29]), which in a number of cases will play the part of positive or negative numbers, respectively. This explains the choice of the figures in (6).

For us the **Minkowski universe** is a set of 49 elements which possess the structure of the two-dimensional vector space M^2 over $\mathcal{F}$ with coordinates $(t, x) \in \mathcal{F} \times \mathcal{F}$ and the product

$$(\mathbf{x}_1, \mathbf{x}_2) = t_1 t_2 - x_1 x_2 \in \mathcal{F} \tag{7}$$

defined in it (in the chosen coordinate system) for any pair of vectors $\mathbf{x}_1 = (x_1, t_1)$, $\mathbf{x}_2 = (x_2, t_2)$. We call the group $\mathcal{L}$ formed by all linear transformations

$$t' = at + bx, \quad x' = ct + dx, \qquad a, b, c, d \in \mathcal{F}, \tag{8}$$

that leave form (7) invariant a *Lorentz group.* We can verify that $\mathcal{L}$ consists of elements Λ_λ which correspond to the transformations for which

$$a = d = 2^{-1}(\lambda + \lambda^{-1}), \quad b = c = 2^{-1}(\lambda - \lambda^{-1}), \qquad \lambda \in \mathcal{F}, \quad \lambda \neq 0,$$

in (8) and of an element π for which $a = 1$, $d = -1$, $b = c = 0$. In particular, the subgroup generated by the elements Λ_λ, $\lambda \in \mathcal{F}_+$ coincides with the so-called commutative subgroup in $\mathcal{L}$ and, by analogy with a continual case, can be called a *proper orthochronous Lorentz group* $\mathcal{L}'$. The addition of the element π to $\mathcal{L}'$ gives the *orthochronous* Lorentz group which is fundamental for further considerations and which consists of six elements [67]. It is denoted by $\mathcal{L}^\uparrow$.

We can visualize the geometry generated by the group $\mathcal{L}^{\uparrow}$ in M^2 if we consider the orbits of the group $\mathcal{L}^{\uparrow}$ in M^2. In particular, we obtain analogs of isotropic straight lines, the direct (V^+) and inverse (V^-) light cones, etc. One of the orbits will play an important part in some constructions that will follow.

The addition of shifts to $\mathcal{L}^{\uparrow}$, i.e., the consideration of inhomogeneous linear transformations together with (8), leads to the Poincaré group $\Pi^{\uparrow}$ with elements denoted by $(\Lambda, \mathbf{a})$, where $\Lambda \in \mathcal{L}^{\uparrow}$ and $\mathbf{a} = (a_0, a_1) \in \mathcal{F} \times \mathcal{F}$ is a shift.

We can now describe the postulates that lie at the basis of the axiomatic approach mentioned at the beginning of the subsection.

We suppose that we can be given the field

$$\Phi : \; M^2 \to \mathfrak{U}(\mathfrak{h}), \qquad \mathbf{x} \mapsto \Phi(\mathbf{x}),$$

where $\mathfrak{h}$ is a complex Hilbert space and $\mathfrak{U}(\mathfrak{h})$ is an algebra of the Hermitian operators. We require that the following axioms be valid.

AXIOM 1. *The unitary representation of the groups* $\Pi^{\uparrow}$ *by the operators*

$$\mathrm{U}(\Lambda, \mathbf{a}) : \; \mathfrak{h} \to \mathfrak{h}, \qquad \Lambda \in \mathcal{L}^{\uparrow}, \quad \mathbf{a} \in M^2,$$

is defined, and

$$\mathrm{U}\Phi\mathrm{U}^{-1} = \Phi(\Lambda\mathbf{x} + \mathbf{a})$$

for the field operator $\Phi(x)$ (**Lorentz covariance**).

In order to formulate the second postulate (the less visual and not used in the sequel), we shall turn to the representation of the group $\Pi^{\uparrow}$ introduced in A.1. In a continual case, the identity operators $\mathrm{U}(1, \mathbf{a})$ (which, in their totality, give the representation of the translation subgroup $T \subset \Pi^{\uparrow}$) can be written ([22, pp. 36, 83]) in the form

$$\mathrm{U}(1, a) = \int e^{i(p,a)}\, dE(p),$$

where p is an element of the space $\tilde{M}^2$ which is the dual (conjugate) of M^2. This representation preserves sense in our model as well, when the integral is replaced by the corresponding sum.

The spectral measure $E(p)$ generated by $\mathrm{U}(1, \mathbf{a})$ allows us to determine the Hermitian operator $P : \; \mathfrak{h} \to \mathfrak{h}$ (of the energy–momentum)

$$P = \int p\, dE(p).$$

It should also be noted that Minkowski's geometry in M^2 induces the corresponding geometry in $\tilde{M}^2$ (see 2.4), making it possible, in particular, to determine the cone $\tilde{V}^+$.

AXIOM 2. *The spectrum* P (the points where $dE(p) \neq 0$) *lies in* $\tilde{V}^+$; *the point* $p = 0$ *belongs to the spectrum; the corresponding proper subspace in* $\mathfrak{h}$ *is one-dimensional and defines vacuum* (the **spectral condition and the stability of vacuum**).

REMARK. We have actually given the continual formulation of the postulate. The more concise, purely algebraic [67] variant has the following form.

AXIOM 2′. The characters of one-dimensional representations of the translation subgroups $T \subset \Pi^\uparrow$, parametrized by the pairs $(p_0, p_1) \in \tilde{M}^2$, belong to $\tilde{V}^+$.

It pays to add the following.

AXIOM 2″. There exists a unique one-dimensional subspace (vacuum) which is invariant relative to all operators $\mathrm{U}(\Lambda, \mathbf{a})$.

Finally, here is the last postulate.

AXIOM 3. *The commutator* $[\Phi(\mathbf{x}), \Phi(\mathbf{x}')] = 0$ *if* $(\mathbf{x}, \mathbf{x}') \in \mathcal{F}_-$ (**localization or causality**).

The axiomatic approach makes it possible to carry out numerous rigorous mathematical constructions but its relations with the "analytic" aspect of QFT, more or less connected with a physical experiment, have not been sufficiently elucidated. In particular, the construction of meaningful models is rather difficult. An idea was recently advanced that for axiomatics to be realized, the space $\mathfrak{h}$ must necessarily possess an indefinite metric (gauge theories).

Below, on the basis of the M^2 that we have constructed, we shall try to model objects that appear in the "analytic" approach.

2.4. Fields operators. Let us construct certain operator fields, $\Phi(\mathbf{x}) : S \to S$, which meet the requirements of covariance and localization that we have mentioned. Let $\tilde{M}$ be a conjugate of the space M (we omit the dimensionality index equal to two), i.e., a space of linear functionals with values in $\mathcal{F}$. Since pairing (7) is nondegenerate, we can introduce in $\tilde{M}$, in an ordinary way, a dual basis such that the values of the functional l_k, that corresponds to the element $\mathbf{k} = (h, k)$, admit of the notation

$$l_{\mathbf{k}} = ht - kx, \tag{9}$$

and the natural isomorphism between $\tilde{M}$ and M induces the pairing

$$(\mathbf{k}_1, \mathbf{k}_2) = h_1 h_2 - k_1 k_2 \tag{10}$$

in $\tilde{M}$. As in the Euclidean case, since pairing (10) is "diagonal", we may not distinguish between covariant and contravariant vectors and assume that the linear transformation Λ in M induces the same transformation in $\tilde{M}$. Thus the transformations from $\mathcal{L}$ leave the values of forms (9), (10) unchanged.

We can isolate in $\tilde{M}$ "mass surfaces", i.e., sets of solutions of the equation

$$h^2 - k^2 = m^2, \qquad m \in \mathcal{F}. \tag{11}$$

In this event, the coordinate h is assigned the meaning of energy and an additional constraint

$$h \in \mathcal{F}_+ \tag{12}$$

is imposed on the solutions of Eq. (11), which is $\mathcal{L}$-invariant, but this requirement is not invariant in our model, generally speaking, even with respect to $\mathcal{L}^\uparrow$. The value $m^2 = 2$ is the only suitable value of m^2 for which the solutions of (11), that satisfy the additional condition (12), form the orbit

$$(4,0), \quad (2,4), \quad (2,-4). \tag{13}$$

The corresponding (entering into (13)) values $k = -4, 0, 4$ will be considered to be "admissible". We also use the fact that every admissible value of k is in a one-to-one correspondence with the point k_j, $j = 1,2,3$, of orbit (13) (this returns us to the situation discussed in 2.2, but it is now inconvenient to use "polynomial" notations).

Let us now construct field operators. In order to construct the unitary representation $\mathrm{U}(\Lambda, \mathbf{a}) : S \to S$ of the group $\Pi^\uparrow$, it is obviously sufficient to define the corresponding representation in S^n for an arbitrary n. If, as it is customary to do, we write the element $|f_{(n)}\rangle \in S^n$, $n \geq 1$ (see 2.2) in the form of the function (symmetric) $f(\mathbf{k}_{(n)}) = f(\mathbf{k}_{j_1}, \ldots, \mathbf{k}_{j_n})$, where j_s assumes values from the set 1, 2, 3, then (cf. [57, p. 164]) we have

$$\mathrm{U}(\Lambda, \mathbf{a}) f(\mathbf{k}_{(n)}) = \exp\left(\nu \sum \mathbf{k}_{j_s} \cdot \mathbf{a}\right) f(\Lambda^{-1}\mathbf{k}_{j_1}, \ldots, \Lambda^{-1}\mathbf{k}_{j_n}), \tag{14}$$

where $\nu = 2\pi i/7$. In this case, $\sum \mathbf{k}_{j_s} \cdot \mathbf{a}$ is regarded as a certain element of $\mathcal{F}$ and $\nu \sum \mathbf{k}_{j_s} \cdot \mathbf{a}$ as a complex number defined by the injection $\mathcal{F} \to \mathbb{Z}$ specified by (6). Relation (14) reflects, in particular, the fact that our basis is the basis of eigenvectors of the translation operator. In S^0 we set $\mathrm{U}(\Lambda, \mathbf{a}) = 1$. From (14) we can get transformation properties of the operators A_j, A_j^* (see 2.2). Using the one-to-one correspondence between the subscripts j and the vectors $\mathbf{k}_j$, we shall write $A_{\mathbf{k}}$. Then we have

$$\mathrm{U}(\Lambda, \mathbf{a}) A_k \mathrm{U}^{-1}(\Lambda, \mathbf{a}) = \exp(-\nu \Lambda \mathbf{k} \cdot \mathbf{a}) A_{\wedge_{\mathbf{k}}}. \tag{15}$$

For $A_{\mathbf{k}}^*$ in (15) we change the sign of the index of the exponent.

Proceeding from $A_{\mathbf{k}}$, we shall define the field operators. We set

$$\Phi_-(t, x) = \exp(-\nu \mathbf{x} \cdot \mathbf{k}_1) A_1 + \exp(-\nu \mathbf{x} \cdot \mathbf{k}_2) A_2 + \exp(-\nu \mathbf{x} \cdot \mathbf{k}_3) A_3,$$

where the exponents are numerical factors. Similarly,

$$\Phi_+(t, x) = \sum_1^3 \exp(\nu \mathbf{x} \cdot \mathbf{k}_j) A_j^*.$$

Let us calculate the commutator of the operators we have constructed that correspond to the points $\mathbf{x}$, $\mathbf{x}' \in M$. We obtain

$$[\Phi_-(\mathbf{x}), \Phi_+(\mathbf{x}')] = \Phi_- \Phi'_+ - \Phi'_+ \Phi_- = \sum \exp(-\nu(\mathbf{x} - \mathbf{x}') \cdot \mathbf{k}_j)[A_j, A_j^*].$$

By virtue of the properties of the operators A_j, this commutator is a numerical function (denoted by $i\Delta_+$) of the difference $\mathbf{x} - \mathbf{x}'$ multiplied by an identity operator. Introducing the notations

$$[\Phi_-(\mathbf{x}), \Phi_+(\mathbf{x}')] = i\Delta_+(\mathbf{x} - \mathbf{x}'),$$

$$[\Phi_+(\mathbf{x}), \Phi_-(\mathbf{x}')] = i\Delta_-(\mathbf{x} - \mathbf{x}'),$$

we immediately get from the definitions the relation

$$\Delta_+(\mathbf{x}) = -\Delta_-(-\mathbf{x}). \tag{16}$$

By virtue of the stipulations that we have made concerning the action of $\mathcal{L}^\uparrow$ in M, $\tilde{M}$, the form $x \cdot k$ is invariant and, consequently, the functions

$$\Delta_\pm(\wedge \mathbf{x}) = \Delta_\pm(\mathbf{x}). \tag{17}$$

that we have introduced are invariant.

Let us now introduce the Hermitian field operator by setting

$$\Phi(t, x) = \Phi_- + \Phi_+.$$

Then

$$[\Phi, \Phi'] = [\Phi_-, \Phi'_+] + [\Phi_+, \Phi'_-] = i(\Delta_+ + \Delta_-)(\mathbf{x} - \mathbf{x}') = i\Delta(\mathbf{x} - \mathbf{x}').$$

However, in accordance with (16),

$$\Delta_+(\mathbf{x}) + \Delta_-(\mathbf{x}) = \Delta_+(\mathbf{x}) - \Delta_-(\mathbf{x}),$$

i.e.,

$$\Delta(\mathbf{x}) = 0 \quad \text{if} \quad \Delta_+(\mathbf{x}) = \Delta_+(-\mathbf{x}),$$

or, by virtue of (17), $\Delta(\mathbf{x})$ vanishes on the set of points $\mathbf{x}$ such that there exists a transformation $\Lambda \in \mathcal{L}^\uparrow$ for which $\Lambda x = -\mathbf{x}$, and this means that $\Delta(\mathbf{x})$ vanishes outside of the interior of the light cone, i.e., the constructed field operator $\Phi(t, x)$ satisfies the localization requirement.

Together with another family of operators $\Psi(t, x)$ ("conjugate momentum"), generated by the second simplest combination of $\Phi_+ \Phi_-$, that yields the self-conjugate operator

$$\Psi(t, x) = i(\Phi_- - \Phi_+),$$

the operators $\Phi(t, x)$ that we have introduced serve as the basic material for dynamic constructions described in the next subsection.

2.5. Scattering and perturbation theory. The physical experiments, the mathematical apparatus for whose description must be given by QFT, are the so-called **scattering** experiments. Let us consider such a description from the point of view of our model.

The initial state of the system at the moment t' is defined by an element $f' \in S$ which is a function $k \in \oplus_0^\infty \tilde{M}$ in the sense of the formalism described in 2.2. In the same way we can describe the finite state f'' at the moment t'' (corresponding to the end of the experiment with scattering). The relationship between these states must be defined by the unitary operator that defines the dynamic picture

$$f'' = \mathrm{U}_{[t', t'']} f'.$$

The probability of discovery of the fixed state f_0 at the moment t'' is defined, in accordance with the laws of the quantum-mechanical description, by the number

$$|\langle f_0 | f'' \rangle|^2 = |\langle f_0 | \mathrm{U}_{[t', t'']} f' \rangle|^2, \tag{18}$$

where the inner product is taken in S.

In a continual case, the initial and final moments are usually set equal to $\pm\infty$ and are not explicitly indicated in relations of the form (18). The operator $\mathrm{U}_{\pm\infty}$ is denoted by S, and, for $f' = f_a$, $f'' = f_b$, the complex number

$$S_{ab} = \langle f_b | S f_a \rangle$$

is called a **matrix element** and the operator itself a S-matrix. It is an open question to what extent is QFT, understood in the sense of constructions from 2.3, 2.4, necessary for constructing the S-matrix. We shall not try to decide it here but shall see at what conclusions we can arrive if we assume that U is defined by the Hermitian operator $\mathbf{H}$, i.e., the Hamiltonian

$$U(t) = \exp(-i\mathbf{H}(t)),$$

which, in turn, can be constructed from the field operators

$$\mathbf{H} = \sum_x \mathbf{F}\{\Phi(\mathbf{x}), \Psi(\mathbf{x})\}, \tag{19}$$

where $\mathbf{F}$ is a function of a simple structure which is chosen on some additional considerations. The Hamiltonian "par excellence", i.e., the energy operator obtained by means of the recalculation of the corresponding operator from 2.2, serves as an important example when we choose $\mathbf{F}$. Even if we abstract ourselves from the magical "correspondence principle" (in the classical field theory this is precisely the way to express the Hamiltonian via the field amplitude, replacing, of course, the sum by an integral), the choice of the representation of $\mathbf{H}$ in the form (19) can be justified by the convenience of accounting, in this assignment, of the requirements of localization and covariance (cf. [4, Sec. 18]). For instance, if

$$\mathbf{H}_g(t) = \sum_x g(\mathbf{x})\mathbf{F}\{\Phi(\mathbf{x}), \Psi(\mathbf{x})\}$$

and the operator $\mathbf{H}_f(t')$ is defined by analogy, and $\mathbf{x} \in \operatorname{supp} g$, $\mathbf{x}' \in \operatorname{supp} f$ imply $|x - x'| > |t - t'|$ (the supports g and f are separated in space), then the operators $\mathbf{H}_g$, $\mathbf{H}_f$ must commute.

The further use of Hamiltonian (19) for our model looks as follows. Suppose that

$$[t', t''] = [\tau_0, \tau_1] + [\tau_1, \tau_2] + \ldots + [\tau_{n-1}, \tau_n],$$
$$\tau_0 = t', \quad \tau_n = t'', \tag{20}$$
$$\mathrm{U}_{[t',t'']} = \exp(-i\mathbf{H}_1) \ldots \exp(-i\mathbf{H}_n),$$

and $\mathbf{H}_k$ defines the evolution on the time interval $[\tau_{k-1}, \tau_k]$. (We imply that the preceding constructions connected with $\mathcal{F}_p$ can be realized for any prime p, and this allows us to consider arbitrarily long chains (20).) We represent $\exp(i\mathbf{H}_k)$ by the series $\sum_s (i\mathbf{H}_k)^s/s!$ and write the corresponding decomposition for U (omitting the explicit indication of the initial and final moments t', t'') as

$$\mathrm{U} = 1 - i\sum_k \mathbf{H}_k - \sum_{k,j} c_{k,j}\mathbf{H}_k\mathbf{H}_j + i\sum_{k,j,s} c_{k,j,s}\mathbf{H}_k\mathbf{H}_j\mathbf{H}_s + \ldots \tag{21}$$

Considering now representation (19) for every $\mathbf{H}_k$, we see that eventually the sum (21) is formed from the combinations of the field operators $\Phi_-(\mathbf{x}_k)$, $\Phi_+(\mathbf{x}_j)$. The corresponding representation for U is known as the **theory of perturbation series**. The name is due to the fact that this construction is correct for the stationary pattern with a time-independent "free Hamiltonian", namely, an energy operator, and is usually used for Hamiltonians obtained from free ones upon the addition of the "interaction" (or "self-action"). In a continual case (even in the simplest models) we do not know of nontrivial examples in which series (21) would converge. Nevertheless, the use of the terms of series (21) is the main technique of constructing concrete (suitable for calculations) matrix elements for relations of form (18).

Let us use a simple model (imaginary, i.e., free of content in a continual case) to consider the formal procedure employed in this case. Let

$$\mathbf{H} = \sum_{\mathbf{x}} \Phi(\mathbf{x}).$$

Then, "in the second order of the perturbation theory", the matrix element S_{ab} has the form

$$\left\langle f_b \middle| \sum_{\mathbf{x}} \sum_{\mathbf{x}'} \Phi(\mathbf{x})\Phi(\mathbf{x}') f_a \right\rangle. \tag{22}$$

Substituting the expressions $\Phi = \Phi_- + \Phi_+$ into (22), we see that we obtain the sums of terms of four types:

$$\Phi_-(\mathbf{x})\Phi_-(\mathbf{x}'), \quad \Phi_-(\mathbf{x})\Phi_+(\mathbf{x}'), \quad \Phi_+(\mathbf{x})\Phi_-(\mathbf{x}'), \quad \Phi_+(\mathbf{x})\Phi_+(\mathbf{x}').$$

We can associate each type with the "Feynman diagram" according to the following law: we mark a pair of points which we denote by $\mathbf{x}$, $\mathbf{x}'$, and associate the factor Φ_- with an arrow entering the corresponding point ("annihilation") and the factor Φ_+ with an arrow emanating from the corresponding point ("birth"). We get the diagrams

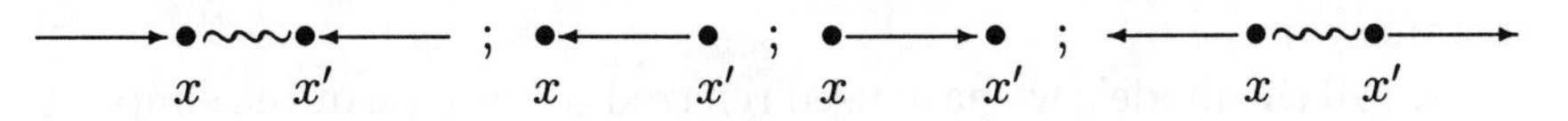

where the wavy line ensures the connectedness of the diagram that corresponds to a certain group of factors. Conversely, when we have appropriate rules, we can associate every diagram with an operator in S defined, in a continual case, by the "Feynman integral". For complicated Hamiltonians such a correspondence that makes it possible to develop specific techniques and intuition proves to be invaluable. Additional laws (for instance, the conservation laws, the considerations of symmetry) allow us immediately to exclude from consideration a number of diagrams. A drawback of the method is the divergence of integrals written on the basis of formal rules.

Returning to (22), we must note that substituting various single-particle amplitudes, say, $f_a(\mathbf{k})$, $f_b(\mathbf{k}')$, into the inner product, we get a complex function of the pair $F(\mathbf{k}, \mathbf{k}')$. In a continual case, this is how numerical and functional characteristics appear that are connected with the experiment with scattering.

2.6. Additional remarks. If we consider the given model as a variant of the outline for QFT, then we must take into account certain additional remarks.

1. The inadequacy of the formal procedure leading to series (21) brought about numerous attempts to construct a scattering theory in which certain properties of the S-matrix are postulated.

Of the utmost importance here is the study and use of the so-called **dispersion relations**. The starting point is the assumption that the S-matrix can be represented as a decomposition with respect to field operators (of the type obtained from (21) upon the substitution of Hamiltonian (19)). The use of the localization properties of field operators and the requirements of the unitary properties and covariance of the S-matrix makes it possible to prove that functions of the form $\mathcal{P}(\mathbf{p}, \mathbf{p}') = \langle f_{\mathbf{p}} | S | f_{\mathbf{p}'} \rangle$ are analytic. The ratio of the real and the imaginary part of $\mathcal{P}$ is known as a dispersion relation and makes it possible to obtain a number of physical corollaries that are related to scattering.

Another variant of the "S-matrix theory" is a direct study of the matrix elements, i.e., analytic functions obtained according to the correspondence recipes: a diagram — Feynman integral ([56], for instance).

2. All the models we have used referred to the case of the simplest scalar field. The actual problems of QFT are connected with the necessity of considering the interaction of fields of different structures, namely, a scalar field, a vector field, a spinor field, an electromagnetic field, etc., i.e., with the consideration of functions over M, $\tilde{M}$, that have several components and can be transformed according to the corresponding representations of the group $\Pi^{\uparrow}$. We can elucidate the principle of constructing Hamiltonians that describe the interaction of fields of different natures, by the way of examples from vector analysis, namely, if we have to determine a bilinear functional of a given scalar field f and a vector field u, then it is natural to consider expressions of the form

$$\int \operatorname{grad} f \cdot u \, dV \quad \text{("vector relationship")}$$

or

$$\int f \cdot \operatorname{div} u \, dV \quad \text{("scalar relationship"), etc.}$$

The Hamiltonians actually used in QFT are very complicated.

3. The analogs of the Klein–Gordon equation (or its generalizations) were absent in our model. In standard expositions of QFT the properties of solutions of these equations are widely used when field operators are introduced (and studied). It is acknowledged, however, that this use is, possibly, only an auxiliary device convenient for obtaining functions that possess specific transformaton properties. Just this point of view is convenient for us.

Chapter 5

Structural Analysis of Discrete Equations

0. Introductory Remarks

The two brief sections of the concluding chapter contain statements of problems rather than the exposition of the results. The first section is devoted to the perturbations of simple "model" systems of equations and the second, to the construction of a discrete analog of jet bundles[1] used in the general theory of solvability of partial differential equations. The corresponding discussions, in a slightly different form, were first published in [17, 18].

As concerns the first section, it should be pointed out that despite its significance and the variety of applications, the perturbation theory for operators in partial derivatives is now in a hardly satisfactory state except for its special divisions [23]. The suggested structural analysis of simple models that refer to the influence exerted by the perturbations of a domain on the spectrum of the operator defined in it may prove to be useful.

The second section originated in the process of the search for an answer to the following question: "To what extent can we use a discrete object for simulating certain concepts used in the general theory of solvability of partial differential equations?" (see [8, 36,

[1] The term "jet" in this context can be associated with the "jet" of splashes corresponding to "cascades" of partial derivatives of different orders of the function f.

45], for instance). The indicated concepts include jet bundles and elements of the homology theory (connected with "diagram technique"). The usefulness of the suggested constructions in concrete problems of a discrete nature is rather problematic.

1. Perturbation of Model Equations

1.0. Preliminary remarks. The content of this section is closely connected with that of Sec. 2, Ch. 0. Model equations are the simplest equations on the discrete circle Ω that contain a translation operator (of the type of Eq. (5), Sec. 2, Ch. 0), or similar equations on the product $\Omega_1 \times \Omega_2$ of these circles. The solutions of these equations (and the spectrum of the corresponding operators) can be immediately calculated either directly or by using the "Fourier transformation". The perturbation of the original structure consists in the addition of a point to Ω (to $\Omega_1 \times \Omega_2$) so that the group of mappings of these sets onto itself, generated by shifts, is no longer Abelian.

The corresponding perturbation is introduced into the model equation and its influence on the spectrum of the resulting "perturbed" operator is considered. The constructions being used admit of many variants and the variant given below may not be the best.

1.1. One-dimensional case. Suppose that Ω is a finite set of N elements labeled by the numbers $k = 1, \ldots, N$, and τ is the map

$$\tau : \Omega \to \Omega, \quad k \mapsto k+1, \quad k \neq N, \quad \tau N = 1.$$

Let H be an N-dimensional Hilbert space of the complex functions $f : \Omega \to \mathbb{C}$ with the inner product

$$(f, g) = \sum_{1}^{N} f(k)\overline{g}(k). \tag{1}$$

We do not introduce normalization in (1) since we do not make any comparisons with a continual case. The shift τ generates, as we have repeatedly pointed out, the unitary operator

$$\hat{\tau} : H \to H, \qquad \hat{\tau} f(k) = f(\tau^{-1} k). \tag{2}$$

The spectrum of the operator $\hat{\tau}$ is defined by the set

$$\lambda_k = e^{-\nu k}, \qquad \nu = 2\pi i N^{-1}, \quad k = 1, \ldots, N$$

of roots of unity and the corresponding eigenfunctions that form the orthogonal basis H have the form

$$\eta_k(x) = e^{\nu k x}, \qquad x = 1, \ldots, k.$$

We form a set Ω' by adding to Ω a point labeled by the number $N+1$ and extend the automorphism τ to Ω' setting $\tau(N+1) = N+1$. At the same time, we introduce an additional automorphism

$$\beta:\ \Omega' \to \Omega', \qquad \beta k = k, \qquad k = 1, \ldots, N-1,$$

$$\beta N = N+1, \qquad \beta(N+1) = N.$$

We denote the Hilbert space of the complex functions over Ω', constructed by analogy with H, by H' and use law (2) to define the unitary operators

$$\hat{\tau},\ \hat{\beta}:\ H' \to H'.$$

Note that the automorphism $\tau\beta$ gives the shift in Ω' similar to the shift τ in Ω. At the same time, $\beta\tau \neq \tau\beta$, i.e., now the group of all automorphisms Ω', generated by the pair τ, β, is no longer commutative. We characterize the spectrum of the operators $\hat{\tau}$, $\widehat{\tau\beta} \equiv \hat{\tau}\hat{\beta}$ noting that it is defined, respectively, by the roots of the equations

$$(\lambda^N - 1)(\lambda - 1) = 0, \qquad \lambda^{N+1} = 0,$$

whose left-hand sides can be regarded as characteristic polynomials corresponding to the ordinary matrix notation of the equations

$$\hat{\tau}u - \lambda u = 0, \qquad \hat{\tau}\beta u - \lambda u = 0.$$

It has been mentioned that there are many ways of introducing perturbations into these simplest equations. Let us consider, for example, the family of operators

$$L_\varepsilon \equiv (1 - \varepsilon)\hat{\tau} + \varepsilon\hat{\tau}\hat{\beta}.$$

The roots of the characteristic polynomial

$$P_\varepsilon(\lambda) = \lambda^{N+1} - (\lambda^N + \lambda - 1)(1 - \varepsilon) - \varepsilon \tag{3}$$

are the eigenvalues of L_ε. However, we cannot now write out the roots and the corresponding eigenfunctions in explicit form, but, nevertheless, we can make certain inferences of a qualitative nature. The following lemma will be useful.

LEMMA. *The mapping of the complex plane*

$$w = \alpha z(\alpha + z)^{-1} \tag{4}$$

for any $\alpha \neq 0$, *lying on the circle* $\Gamma: \; |z-1|$, *maps* Γ *into the circle* $\gamma: \; |w - 1/2| = 1/2$.

PROOF. We shall note, first of all, that the image of the real axis under mapping (4) is either a circle that touches the real axis at the point $w = 0$ (for $\operatorname{Im} \alpha \neq 0$) or the real axis ($\alpha = 2$). The circle γ (the image of Γ) cuts the real axis at the points $w = 0$ ($z = 0$) and $w = 1$. In order to verify the last relation, it is sufficient to set $z = \overline{\alpha}$, $\alpha = 1 - e^{i\beta}$. At the same time, since the mapping (3) is conformal, γ is orthogonal at zero to the image of the real axis and, consequently, according to the remark we have made, to the real axis itself. We infer that the point $w = 1/2$ is the center of γ. ■

PROPOSITION 1. *When* ε *varies from* 0 *to* 1 *in the lower half-plane along the hemisphere*

$$2\varepsilon(x) = \exp[i\pi(1-x)] + 1, \qquad 0 \leq x \leq 1,$$

the roots $\lambda_k(\varepsilon)$, $k = 0, \ldots, N-1$, *of polynomial* (3) *are moving counterclockwise along a unit circle from the values* $\lambda_k(0) = e^{\nu k}$ *to the values* $\lambda_k(1) = e^{\mu(k+1)}$, $\mu = 2\pi i(N+1)^{-1}$. *The root* $\lambda_N = 1$ *remains fixed.*

PROOF. The equation $P_\varepsilon(\lambda) = 0$ allows us to express ε as a function of λ, i.e.,

$$\varepsilon = (1-\lambda)(1-\lambda^N)[(1-\lambda) + (1-\lambda^N)]^{-1}, \qquad \lambda \neq 1$$

(for $\lambda = 1$ we must perform the necessary canceling). It remains to be noted that the displacement of the point $\lambda_k(\varepsilon)$ from $\lambda_k(0)$ to $\lambda_k(1)$ along a unit circle corresponds to the motion of α and z in (4) along the circle Γ and to use the lemma that we have proved. ■

It pays to consider the special basis in H' which is convenient for studying some characteristics of the operator L_ε [18]. Let us define the vectors $\vartheta_k(x)$, $k = 1, \ldots, N$, that coincide with $\eta_k(x)$ for $x \neq N+1$ and are complemented by the component $\vartheta_k(N+1) = 0$.

Then we introduce the vector $\vartheta_0(x)$ with the only nonzero component $\vartheta_0(N+1) = 0$. We take ϑ_0, $\vartheta_k + \vartheta_0$ as the basis. Then

$$\hat{\tau}\hat{\beta}(\vartheta_k + \vartheta_0) = \hat{\tau}(\vartheta_k + \vartheta_0) = \lambda_k \vartheta_k + \vartheta_0 = \lambda_k(\vartheta_k + \vartheta_0) + (1 - \lambda_k)\vartheta_0,$$

whence it follows that

$$\hat{\tau}\hat{\beta}\vartheta_0 = N^{-1}\sum_1^N \lambda_k(\vartheta_k + \vartheta_0), \qquad \lambda_k = e^{\nu k}.$$

The basis we have constructed makes it possible, in particular, to obtain a different form of notation of polynomial (3). We shall present it for a more complicated case in 1.2. For the time being, observe that the notation found in this way is convenient for calculating the derivative

$$\frac{d\lambda}{d\varepsilon}\Big|_{\substack{\lambda=\lambda_k \\ \varepsilon=0}} = -\lambda_k N^{-1}$$

(it stands to reason that the fixed root is excluded).

Derivatives of this kind often contain useful information.

1.2. Two-dimensional case.

The part of the set Ω is now played by

$$\Omega = \Omega_1 \times \Omega_2, \quad \Omega_1 = \{1, \dots, N\}, \quad \Omega_2 = \{1, \dots, M\}.$$

The pair of commuting automorphisms τ, χ is defined on Ω. For τ we have

$$\tau: \Omega_1 \to \Omega_1, \quad k \mapsto k+1, \quad k \neq N, \quad \tau N = 1, \quad \tau \,|\, \Omega_2 = 1;$$

and the action $\chi: \Omega \to \Omega$ can be defined by analogy (a shift on Ω_2, a unity on Ω_1). The space of the complex functions over Ω is an NM-dimensional Hilbert space $H = H_1 \otimes H_2$, where H_σ is the corresponding space over Ω_σ, $\sigma = 1, 2$. The elements $\eta_{k,s}(x, y)$

$$\begin{aligned} \eta_{k,s} &= \eta_k(x)\eta_s(y), \qquad & \eta_k(x) &= e^{\nu k x}, \\ \nu &= 2\pi i N^{-1}, \qquad & \eta_s(y) &= e^{\mu s y}, \end{aligned} \tag{5}$$

form an orthogonal basis of eigenvectors of the operators $\hat{\tau}$, $\hat{\chi}$ generated (in accordance with (2)) by the automorphisms that we have introduced. Operators of the form

$$L = q\hat{\tau} + r\hat{\chi}, \qquad q, r = \text{const} \tag{6}$$

are model operators.

Let us introduce a set $\tilde{\Omega}$ by adding to Ω a point, which we denote by ω. Then we extend the automorphisms τ, χ to $\tilde{\Omega}$ assuming that they leave the added point fixed, and define the automorphism β setting

$$\beta: \omega \mapsto (M, N), \qquad (M, N) \mapsto \omega.$$

It leaves the other points of $\tilde{\Omega}$ fixed.

Just as in a one-dimensional case, there are many variants of the "perturbation" of the operator (6) which are due to the use of $\hat{\beta}$ that violated the commutability of the original structure. We shall discuss the variant which is the closest to that considered in 1.1. Assuming that $q = 1$ in (6), we set

$$L_\varepsilon = (1 - \varepsilon)L + \varepsilon L\hat{\beta}. \tag{7}$$

It is difficult to write out for (7) the characteristic polynomial in the form similar to (3). We use the extension of basis (5) of the type suggested in 1.1.

We retain the notation $\eta_{k,s}$ for the elements obtained from the respective elements (5) by adding zero as the last component and denote by η_0 the element whose only nonzero component is unity on the $(NM + 1)$th place. We take

$$\eta_{k,s} + \eta_0, \eta_0, \qquad k = 1, \ldots, N, \quad s = 1, \ldots, M$$

as the basis. Setting $\lambda_k = e^{\nu k}$, $\zeta_s = e^{\mu s}$, we have

$$L\hat{\beta}(\eta_{k,s} + \eta_0) = (\lambda_k + r\xi_s)(\eta_{k,s} + \eta_0) + [1 - \lambda_k + r(1 - \xi_s)]\eta_0,$$

$$L\eta_0 = (1 + r)\eta_0,$$

$$L\hat{\beta}\eta_0 = M^{-1}N^{-1}\sum_{k,s}(\lambda_k + r\zeta_s)(\eta_{k,s} + \eta_0).$$

The characteristic polynomial can now be represented as

$$\mathcal{P}_\varepsilon(\lambda) = [(1 + r)(1 - \varepsilon) - \lambda]Q -$$

$$\varepsilon M^{-1}N^{-1}\sum_{k,s}(\lambda_k + r\zeta_s)[1 - \lambda_k + r(1 - \zeta_s)]Q_{k,s},$$

$$Q = \prod_{k,s=1}^{N,M}(\lambda_k + r\zeta_s - \lambda),$$

and $Q_{k,s}$ results from Q upon the deletion of the factor with the corresponding subscripts. Just as in 1.1, we can now calculate

$$\frac{d\lambda}{d\varepsilon}\Big|_{\substack{\lambda=\lambda_k+r\xi_s\\ \varepsilon=0}} = -\lambda_k(MN)^{-1}.$$

This derivative gives some information concerning the nature of the "shift" of the eigenvalues upon the perturbation, and we shall restrict our discussion to this observation.

2. The Formal Theory of Solvability

As has been stipulated, the construction suggested in this section will hardly be useful in the analysis of solvability of concrete systems of difference equations (which are a special case of the model equations under consideration). Nevertheless, this construction is interesting from the methodological point of view.

The concepts of the elementary homology theory that we use go beyond the framework of the information given in Sec. 2, Ch. 1. It is useful to complement them (in the part referring to the exact homological sequence) by the acquaintance with the introductory chapters from [21, 31]. The independent exposition of the "diagram technique" suitable for the use in the situation being considered can be found in Sec. 2, Ch. 1 of [36].

Suppose that $\mathbb{Z}^n$ is the set of points $x = (x_1, \ldots, x_n)$ with integral coordinates (we can consider it to be a zero-dimensional skeleton of our model of the Euclidean space of Ch. 3) and $X \subset \mathbb{Z}^n$ is a subset which we suppose to be finite. The direct product $X \times \mathbb{R}^n = E(X)$ is the *fibration* over X and the function

$$u: X \to \mathbb{R}^n, \qquad x \mapsto u(x)$$

is the *section* of this fiber bundle. We are interested in the general linear first-order model equations over X, i.e., equations (systems) written as

$$a_r^{ip}(x)\tau_i u_p(x) = f_r(x), \qquad x \in X, \tag{L}$$
$$i = 0, \ldots, n, \qquad p = 1, \ldots, N, \qquad r = 1, \ldots, M,$$

where τ_i is a shift along the coordinate x_i: $\tau_i x = (x_1, \ldots, x_i + 1, \ldots, x_n)$, $i = 1, \ldots, n$, $\tau_i u_p = u_p(\tau_i x)$, and τ_0 is an identity operator. The summation is supposed to be carried out over the repeating subscripts. Relation (L) has sense for any $x \in X$ if and only if the section $u(x)$ is defined over a certain "neighborhood" $\tilde{X} \supset X$ which must be described. Suppose that $\tau_i X$ is a set obtained from X by the application of the shift τ_i to every point $x \in X$. Then

$$\tau X = \bigcup_i \tau_i X, \qquad \tilde{X} = X \cup \tau X, \qquad bX = \tilde{X} \setminus X.$$

The set bX will play the part of the boundary of X.

In what follows, we set $M = N$ in (L) and construct a certain special formalism in the framework of which we shall give the description of the character of solvability of system (L) and then, having fixed a certain subset $Y \subset X$, consider the relationship between the solvability of (L) over X, over $X \setminus Y$, and over Y. Let us consider a trivial example illustrating the statement of the problem of interest to us.

EXAMPLE. Suppose that X is a triple of points 0, 1, 2 and the points 1, 2 form Y. Let $u: X \to \mathbb{R}^1$ and (L) have the form

$$0 \cdot u(0) + \tau u(0) = 0, \qquad u(1) + \tau u(1) = 1,$$

$$u(2) + 0 \cdot \tau u(2) = 0.$$

The system is solvable over $X \setminus Y$ and over Y, but is unsolvable over X.

Thus we note that for $M = N$ the equations (L) define the map of the sections

$$L: \Gamma E(\tilde{X}) \to \Gamma E(X). \tag{1}$$

At the same time, together with $E(X)$, we shall consider the jet-fibration $J(X) = X \times \mathbb{R}^{N(n+1)}$ and the mapping

$$j: \Gamma E(\tilde{X}) \to J(X), \qquad u \mapsto \xi,$$

$$\xi_{ip}(x) = \tau_i u_p(x), \qquad i = 0, \ldots, n, \quad p = 1, \ldots, N.$$

As a result, map (1) can be associated with the commutative diagram

$$\begin{array}{ccc} \Gamma E(\tilde{X}) & \xrightarrow{j} & \Gamma J(X) \\ L\downarrow & & \mathcal{L}\downarrow \\ \Gamma E(X) & \xrightarrow{1} & \Gamma E(X), \end{array}$$

that defines the **map of the fibrations**

$$\mathcal{L}: J(X) \to E(X),$$

i.e., the **homomorphism** (linear mapping) of every one of the **linear spaces** $J(x)$, $E(x)$, $x \in X$.

REMARK. The mapping of fibrations always induces a certain **map of sections** but not vice versa, as is shown by the example of the maps of J, L. The construction being presented precisely gives the transition from the map of sections to homomorphisms of fibrations.

We can always identify $E(X)$ with the subfibration $J^0(X) \subset J(x)$ defined by the vectors $\{\xi_{0p}(x)\}$. We shall use this subfibration when we define one more object, namely, the fibration $J^0(X) \otimes T^n$ which associates the point $x \in X$ with the nN-dimensional space of vectors $\{\alpha_{sp}(x)\}$ (an analog of a tangent fibration). We introduce the mapping

$$\begin{gathered} D: \ J(x) \otimes J^0(bx) \to J^0(x) \otimes T^n, \qquad \xi \to \alpha, \\ \alpha_{sp}(x) = \xi_{sp}(x) - \xi_{0p}(\tau_s x). \end{gathered} \tag{2}$$

Although (2) defines D as a map of sections, in the present case it defines a map of fibrations. The following statement elucidates the purpose of introducing D.

PROPOSITION 1. *The relation $D\xi = 0$ is satisfied if and only if $\xi = ju$, where $u \in \Gamma E(X)$.*

PROOF. If $\xi = ju$, then the relation $D\xi = 0$ follows from (2). Suppose now that $\xi \in \operatorname{Ker} D$. We set $u_p(x) = \xi_{0p}$ and denote ju by η. We obtain

$$\eta_{ip}(x) = \tau_i u_p(x) = u_p(\tau_i x) = \xi_{0p}(\tau_i x) = \xi_{ip}(x),$$

i.e., $\eta = \xi$. ■

COROLLARY. *The sequence of homomorphisms*

$$0 \to \Gamma E(\tilde{X}) \xrightarrow{j \oplus j^0} J(X) \oplus J^0(bX) \xrightarrow{D} J^0(X) \otimes T^n \to 0 \tag{3}$$

is exact, i.e., the image of the preceding homomorphism exactly coincides with the kernel of the succeeding one. ■

We can associate (3) with the one-dimensional complex

$$K(X) = K^1(X) \oplus K^0(X),$$

regarding as the chains K^1, K^0 the sections of the fibrations defined by the third and the fourth term, respectively, of the sequence (3) and D as a boundary operator.

REMARK. The homology of the complex we have introduced is of no interest since it is obvious that $\mathcal{H}^1(K)$ is a free Abelian group with the number of generators $N \operatorname{mes} \tilde{X}$ ($\operatorname{mes} \tilde{X}$ is the number of points) and $\mathcal{H}^0(K) = 0$. However, we shall need the complex itself.

In order to complete the constructions, we shall introduce a trivial zero-dimensional complex $F(X)$ over X, whose chains are the sections $\Gamma E(X)$, and extend the map $\mathcal{L}$ to K^1 setting

$$
\begin{aligned}
&\mathcal{L}:\ J(X) \to F(X), \qquad \xi \mapsto f, \qquad f_q(x) = a_q^{ip}(x)\xi_{ip}(x),\\
&\mathcal{L}:\ J^0(bX) \to 0.
\end{aligned}
$$

We introduce the notation $\operatorname{Ker} \mathcal{L} = R^1 \subset K^1$, and represent the final result as a commutative diagram

$$
\begin{array}{ccccccc}
 & & 0 & & 0 & & \\
 & & \downarrow & & \downarrow & & \\
 & & R'(X)\oplus J^0(bX) & \xrightarrow{D} & J^0(X)\oplus T^n & \longrightarrow & 0\\
 & & \downarrow{\scriptstyle \varkappa\oplus 1} & & \downarrow{\scriptstyle 1} & & \\
0\longrightarrow \Gamma E(\tilde{X}) & \xrightarrow{j\oplus j^0} & J(X)\oplus J^0(bX) & \xrightarrow{D} & J^0(X)\oplus T^n & \longrightarrow & 0\\
\downarrow{\scriptstyle L} & & \downarrow{\scriptstyle \mathcal{L}\oplus 0} & & \downarrow & & \\
0\longrightarrow \Gamma E(X) & \xrightarrow{1\oplus 0} & F(X)\oplus 0 & \longrightarrow & 0 & & \\
\downarrow & & \downarrow & & & & \\
0 & & 0 & & & &
\end{array}
$$

in which $\varkappa:\ R^1 \to J$ is a certain fixed homomorphism that defines the isomorphism $R^1 = \operatorname{Ker} \mathcal{L}$ (the parametrization of the kernel) and the operator D in the upper row is defined by the superposition $D \circ (\varkappa \oplus 1)$.

It follows from the definitions that the exactness of the sequences appearing in the diagram can be violated only in the extreme left column, in $\Gamma E(X)$, and in the upper row, in the term $J^0 \otimes T^n$. We associate the upper row with the complex

$$
Q = Q^1(X) \oplus Q^0(X)
$$

with the boundary operator D. The technique of the use of the diagram (and the purpose of its application) is given in the following statement [36].

PROPOSITION 2. *The relations*

$$\dim \mathcal{H}^1(Q) = \dim \operatorname{Ker} L,$$

$$\dim \mathcal{H}^0(Q) = \dim \operatorname{Coker} L$$

hold true. ■

EXAMPLE. We shall illustrate the presented scheme in a completely trivial situation. Suppose that X consists of the only point 1, τX consists of the point 2, and u has a single component. We have

$$a^1 \tau u(1) + a^0 u(1) = f(1), \tag{L}$$

$$a^1 \xi_1(1) + a^0 \xi_0(1) = f(1), \tag{$\mathcal{L}$}$$

$$D\xi(1) = \xi_1(1) - \xi_0(\tau \cdot 1) = \xi_1(1) - \xi_0(2) = \alpha_1(1).$$

Passing to the bases $\{e\}$, $\{h\}$ in K^1, K^0, we obtain

$$\xi = \xi_0(1)e^0(1) + \xi_1(1)e^1(1) + \xi_0(2)e^0(2),$$

$$\alpha = \alpha_1(1)h, \qquad h \equiv h^1(1),$$

$$D\xi(1) = \xi_1(1)De^1(1) + \xi_0(2)De^0(2) = (\xi_1(1) - \xi_0(2))h,$$

whence we have

$$De^1(1) = h, \qquad De^0(2) = -h, \qquad De^0(1) = 0.$$

Let $|a^1| + |a^0| \neq 0$. Then $\operatorname{Ker} \mathcal{L} = R^1$ is one-dimensional and we can introduce a basis $Q^1 = R^1 \oplus J^0(bX)$ of the form $\varepsilon(1)$, $\varepsilon(2)$; we retain the notation h for the base element Q^0,

$$\varkappa : \ \varepsilon(1) \mapsto \varkappa_1 e^1(1) + \varkappa_0 e^0(1) = a^0 e^1(1) - a^1 e^0(1) \in \operatorname{Ker} \mathcal{L},$$

$$1 : \ \varepsilon(2) \mapsto e^0(2)$$

(we do not write ta^0, ta^1 for we wish to have a fixed homomorphism $\varkappa$),

$$D\varepsilon(1) = D \circ \varkappa\varepsilon(1) = a^0 h, \qquad D\varepsilon(2) = D \circ 1\varepsilon(2) = -h.$$

Thus $\varepsilon(1) + a_0\varepsilon(2)$, which is a base cycle in $\mathcal{H}^1(Q)$, is the only nonzero element of $\mathcal{H}(Q)$ corresponding to the one-dimensional kernel of L.

Such a direct calculation in examples in less trivial situations (in particular, in situations connected with the consideration of the partition of X into $Y \subset X$ and $X \setminus Y$ given below) does not cause essential difficulties but is very cumbersome.

Thus, suppose now that $Y \subset X$ is a subset of X. In turn, constructions connected with (L), similar to those carried out for X, can be carried out for Y. In particular, the complex $K^1(Y) \oplus K^0(Y) = K(Y)$ is defined. It is an *open* subcomplex $K(X)$ in the sense that the subcomplex $K[Z] = K(X) \setminus K(Y)$, connected with the set $Z = X \setminus Y$ is closed. We understand the closeness as the permutability of the imbedding operation $I: \ K[Z] \to K(X)$ with the operation D. The brackets in the notation $K[Z]$ indicate that this complex, connected with the set Z, has been constructed according to the recipe different from that used for defining $K(X)$, $K(Y)$.

We can also construct a one-dimensional complex $Q(Y)$ for Y. In this case, it is convenient to suppose that the homomorphisms $\varkappa$ have been chosen once and for all for every $x \in X$. Let us also consider a subcomplex $Q\{Y\}$ in $Q(X)$ generated by the elements (compare the example) entering into $Q(Y)$. In what follows we suppose the system (L) to be *nondegenerate*, and this means that the relation $Q\{Y\} = Q(Y)$ is satisfied.

As in the discussion carried out above for K, we define the complex $Q[Z] = Q(X) \setminus Q(Y)$, which is a closed subcomplex of $Q(X)$. As a result, we obtain an exact sequence of one-dimensional complexes

$$0 \to Q[Z] \xrightarrow{I} Q(X) \xrightarrow{\pi} Q(Y) \to 0, \tag{4}$$

where I is an injection and π is a projection. The exact **homological sequence**

$$0 \to \mathcal{H}^1_Q[Z] \to \mathcal{H}^1_Q(X) \to \mathcal{H}^1_Q(Y) \to \mathcal{H}^0_Q[Z] \to \mathcal{H}^0_Q(X) \to \mathcal{H}^0_Q(Y) \to 0,$$

that contains, according to Proposition 2, some information concerning the solvability of (L) over X in terms with the solvability of the corresponding systems over Y and over Z, is connected with (4) in a standard way [21, 31].

The construction we have presented makes it possible to connect with (L) the so-called **Meyer–Vietoris sequence** [21] that contains information on the solvability of (L) over $X \cup Y$ $(Y \not\subset X)$ according to the properties of the system over X, Y, and $Z = X \cap Y$.

This completes the modeling of the methods that make it possible to connect the solvability theory with the homological algebra.

References

1. J. Adams, *Lectures on the Lie Groups*, W.H. Benjamin, New York–Amsterdam (1969).

2. P. S. Alexandrov, "On the concept of a space in topology," *Usp. Mat. Nauk*, **2**, Issue 1 (1947), 5–57.

3. A. V. Bitsadze,"A space analog of the integral of the Cauchy type and some of its applications," *Izv. Akad. Nauk SSSR, Ser. Mat.*, **17**, No. 6 (1953), 525–538.

4. N. N. Bogolyubov and D. V. Shirkov, *Introduction to the Theory of Quantized Fields* [in Russian], Nauka, Moscow (1984).

5. R. Baxter, *Exactly Solved Models in Statistical Mechanics*, Academic Press, London (1982).

6. H. Weyl, "Mathematics. Theoretical physics," In: *Selected Works* [Russian translation], Nauka, Moscow (1984).

7. N. Ya. Vilenkin, *Special Functions and the Theory of Group Representation* [in Russian], Nauka, Moscow (1965).

8. A. M. Vinogradov, I. S. Krasil'shchikov, and V. V. Lychagin, *Introduction to the Geometry of Nonlinear Differential Equations* [in Russian], Nauka, Moscow (1986).

9. M. Goldberger and K. Watson, *Collision Theory*, John Wiley & Sons, New York (1964).

10. N. Danford and J. Schwartz, "Linear operators. 1," *General Theory*, Interscience Publishers, New York (1958).

11. N. Danford and J. Schwartz, "Linear operators. 2," *Spectral Theory*, Interscience Publishers, New York (1963).

12. A. A. Dezin, "Boundary value problems for certain first-order symmetric linear systems," *Mat. Sb.*, **49 (91)**, No. 4 (1959), 459–484.

13. A. A. Dezin, "Theorems on the existence and uniqueness of solutions of boundary value problems for partial differential equations in functional spaces," *Usp. Mat. Nauk*, **14**, No. 3 (1959), 21–73.

14. A. A. Dezin, "Invariant differential operators and boundary value problems," *Trudy MIAN*, **68**, USSR Acad. Sci. Publ., Moscow (1962).

15. A. A. Dezin, "On the method of orthogonal decompositions," *Sib. Mat. Zh.*, **9**, No. 5 (1968), 1062–1074.

16. A. A. Dezin, "Certain model connected with Euler's equations," *Differ. Uravn.*, **6**, No. 1 (1970), 17–26.

17. A. A. Dezin, "Global solvability of multidimensional difference equations and a construction of the type of the Spencer construction," *Dokl. Akad. Nauk SSSR*, **196**, No. 1 (1971), 28–31.

18. A. A. Dezin, "On the spectrum of some difference operators," *Sib. Mat. Zh.*, **13**, No. 1 (1972), 86–93.

19. A. A. Dezin, *Partial Differential Equations*, Springer, Heidelberg (1987).

20. A. A. Dezin, "Invariant forms and some structural properties of Euler's hydrodynamic equations," *Z. Anal. Anwend.*, **2**, No. 5 (1983), 401–409.

21. A. Dold, *Lectures on Algebraic Topology*, Springer, Heidelberg (1972).

22. R. Jost, *The General Theory of Quantized Fields*, Amer. Math. Soc., Providence, RI (1965).

23. T. Kato, *Perturbation Theory for Linear Operators*, Springer, Heidelberg (1966).

24. J. Kelley, *General Topology*, Van Nostrand, Princeton (1955).

25. A. A. Kirillov, *Elements of the Representation Theory* [in Russian], Nauka, Moscow (1972).

26. A. I. Kostrikin, *Introduction to Algebra* [in Russian], Nauka, Moscow (1977).

27. A. I. Kostrikin and Yu. I. Manin, *Linear Algebra and Geometry* [in Russian], Nauka, Moscow (1986).

28. N. E. Kochin, I. A. Kibel, and I. V. Roze, *Theoretical Hydrodynamics* [in Russian], Fizmatgiz, Moscow (1963).

29. S. Lang, *Algebra*, Addison–Wesley, Reading, MA (1965).

30. S. Lefschetz, *Algebraic Topology*, Princeton, Princeton, NJ (1942).

31. S. Maclane, *Homology*, Springer-Verlag, W. Berlin (1963).

32. V. P. Maslov, *Perturbation Theory and Asymptotic Methods*, Moscow State University Press., Moscow (1965).

33. J. Milnor and J. Stasheff, *Characteristic Classes*, University Press., Princeton, NJ (1974).

34. T. H. Milne Thomson, *Theoretical Hydrodynamics*, Macmillan & Co., London (1960).

35. A. F. Nikiforov, S. K. Suslov, and V. B. Uvarov, *Classical Orthogonal Polynomials in a Discrete Variable* [in Russian], Nauka, Moscow (1985).

36. V. P. Palamodov, *Linear Differential Operators with Constant Coefficients* [in Russian], Nauka, Moscow (1967).

37. B. V. Pal'tsev, "Multidimensional analog of the Morera theorem," *Sib. Mat. Zh.*, **4**, No. 6 (1963), 1376–1388.

38. M. M. Postnikov, "The homology theory of smooth manifolds and its generalizations," *Usp. Mat. Nauk*, **11**, No. 1 (1956), 115–166.

39. J. de Rham, *Varietés Différentiables*, Hermann & Co., Paris (1955).

40. S. Saks, *Theory of the Integral*, G. E. Stechert Co., New York, USA (1939).

41. A. A. Samarsky, V. F. Tishkin, A. P. Favorsky, and M. Yu. Shashkov, "Operator difference schemes," *Differ. Uravn.*, **17**, No. 7 (1981), 1317–1327.

42. J. Serre, *Représentations Lineares des Groupes Finits*, Hermann & Co., Paris (1967).

43. S. L. Sobolev, *Some Applications of Functional Analysis in Mathematical Physics* [in Russian], Leningrad State University Press, Leningrad (1950).

44. S. L. Sobolev, "On one new problem of mathematical physics," *Izv. Akad. Nauk SSSR, Ser. Mat.*, **18**, No. 1 (1954), 3–50.

45. D. Spencer, "Overdetermined systems of linear partial differential equations," *Bull. Amer. Math. Soc.*, *75*, No. 2 (1969), 179–239.

46. M. Spivak, *Calculus on Manifolds*, W. A. Benjamin, New York (1965).

47. S. Sternberg, *Lectures on Differential Geometry*, Prentice-Hall, Englewood Cliffs, NJ (1964).

48. N. Steenrod, *The Topology of Fiber Bundles*, Princeton U.P., Princeton, NJ (1951).

49. O. V. Troshkin, "On some properties of Euler's fields," *Differ. Uravn.*, **18**, No. 1 (1982), 138–144.

50. H. Whitney, *Geometric Integration Theory*, Princeton U.P., Princeton, NJ (1957).

51. Fam Ngok Thao, "Boundary value problems for natural differential operators on manifolds," *Differ. Uravn.*, **6**, No. 5 (1970), 877–888.

52. D. K. Faddeev, *Lectures on Algebra* [in Russian], Nauka, Moscow (1984).

53. H. Federer, *Geometric Measure Theory*, Springer-Verlag, Berlin–New York (1969).

54. V. G. Fedorovsky, "Estimates of the solutions of first-order invariant hyperbolic systems," *Sib. Mat. Zh.*, **12**, No. 1 (1971), 197–203.

55. P. Halmos, *Measure Theory*, D. Van Nostrand Co., Princeton N.J. (1950).

56. R. Hwa and V. Teplitz, *Homology and Feyman Integrals*, W. A. Benjamin, New York (1966).

57. S. Schweber, *An Introduction to Relativistic Quantum Field Theory*, Row Peterson & Co., New York (1961).

58. B. Schutz, *Geometrical Methods of Mathematical Physics*, Cambridge Univ. Press, Cambridge (1980).

59. V. Arnold, "Sur la geometrie differentielle de groupes de Lie de dimension infinite et ses applications a l'hydrodynamique de fluides parfaits," *Ann. Inst. Fourier*, **16**, No. 1 (1966), 319–361.

60. M. Berger, P. Gauduchon, and E. Mazet, *Le spectre d'une variete rimannienne*, Springer-Verlag, Berlin (1971) (Lect. Notes, Math. **194**.)

61. G. Duff and D. Spencer, "Harmonic tensors on Riemannian manifolds with boundary," *Ann. Math.*, **56**, No. 1 (1952), 128–156.

62. K. Friedrichs, "Symmetric hyperbolic linear differential equations," *Commun. Pure Appl. Math.*, **7**, No. 2 (1954), 345–392.

63. K. Friedrichs, "Differential forms on Riemannian manifolds," *Commun. Pure Appl. Math.*, **8**, No. 4 (1955), 551–590.

64. K. Friedrichs, "Symmetric positive linear differential equations," *Commun. Pure Appl. Math.*, **11**, No. 3 (1958), 333–418.

65. W. Hodge, *The Theory and Applications of Harmonic Integrals*, Cambridge University Press, Cambridge (1952).

66. G. Moisil and N. Theodorescu, "Fonctions holomorphes dans l'espace," *Mathematica*, **5** (1931), 142–159.

67. H. Joos, "Group-theoretical models in local field theories," *J. Math. Phys.*, **5**, No. 2 (1964), 155–164.

68. A. A. Dezin, "Certain models connected with Yang–Mills equations," *Differ. Uravn.*, **29**, No. 5 (1993), 846–851.

69. A. A. Dezin, "On Schrödinger and Klein–Gordon equations," *Differ. Uravn.*, **30**, No. 6 (1994), 1023–1033.

Index